K

After Math

DATE DUE

JAN 1 4 1997	
JAN 2 2 1997	
MAR 2 1 1997 /29	
JUN 3 0 1997. OCT 1 7 1997	
OCT 2 8 1997	
FEB 1 3 1998	
FEB 2 3 1998 SEP 2 8 1998	
SEP 2 7 1999	
JAN 2 6 2001 OCT - 1 2001	
DEC 1 9 2001	

To Judy, Dan, and Paul

Ed Barbeau is professor of mathematics at the University of Toronto. For a long time, he has been interested in conveying mathematics to a wide audience. For several years, he ran the Metro Mathematics Club, founded by W. W. Sawyer, for Toronto high school students, and has given courses for mathematical amateurs at the University of Toronto's School of Continuing Studies. In 1982, he presented a three-part series on "Proof and Truth in Mathematics" for the Canadian Broadcasting Corporation radio series, *Ideas*, and more recently has been heard several times on the popular science CBC broadcast, *Quirks* and *Quarks*. In 1990, he addressed the Royal Canadian Institute on the topic, "How Miraculous is Mathematics?" In 1991, he received a David Hilbert Award from the World Federation of National Mathematics Competitions for contributions to the enrichment of mathematics learning through the stimulation of mathematics challenges. He is a Fellow of the Ontario Institute for Studies in Education.

After Math

Puzzles and Brainteasers

Ed Barbeau

Wall & Emerson, Inc.
Toronto, Ontario • Dayton, Ohio

Requests for permission to make copies of any part of this work should be sent to: Wall & Emerson, Inc., Six O'Connor Drive, Toronto, Ontario, Canada M4K 2K1

Orders for this book may be directed to:

Wall & Emerson, Inc.   *or*   Wall & Emerson, Inc.
Six O'Connor Drive   8701 Slagle Rd.
Toronto, Ontario, Canada   Dayton, Ohio 45458
M4K 2K1

Or by telephone or facsimile:

Telephone: (416) 467-8685 Fax: (416) 696-2460

Canadian Cataloguing in Publication Data

Barbeau, Edward, 1938–
   After math : puzzles and brainteasers

Includes index.
ISBN 0-921332-42-4

1. Mathematical recreations. I. Title.

QA95.B37 1995   793.7'4   C95-930869–5

ISBN 0-921332-42-4
Printed in Canada by Best Book Manufacturers Inc.

1 2 3 4 5 6   04 03 02 01 00 99 98 97 96 95

Table of Contents

Preface

In 1984, the editors of the University of Toronto *Alumni Magazine* asked me to produce a mathematics problems column. As it appeared at the end of each issue, with the Cryptic Crossword, it bore the title "Aftermath." The audience consisted of people who, while university graduates, may not have had mathematics beyond high school, and it was a challenge to come up with problems that were challenging and conveyed something of the flavour of the different facets of mathematics, without requiring advanced knowledge. During its nine years of appearance, I heard from several hundred people in many walks of life and about two dozen became regular correspondents. For them, I circulated a newsletter for the exchange of solutions, ideas, and additional problems. This book includes material from both the *Alumni Magazine* and the newsletters.

Among those respondents who had not touched mathematics in years, some came up with neat solutions. Indeed, it seemed that lack of background could be an advantage; not having access to standard techniques, they had to rely on their native ability and creativity.

Some readers may be into Cryptic Crosswords. My wife and I have recently begun doing these. The mental processes involved seem to be similar in some ways to those used in solving a mathematical problem. Often the answer does not come right away, and one has to indulge in a little free association and piecing together of clues. It is frequently necessary to "sleep on it." But in the end, when the solution comes, it is an occasion for satisfaction and appreciation of the richness of the puzzle. However, the audiences for mathematical puzzles and for Cryptic Crosswords may overlap, but they surely do not coincide. Everyone to their own taste!

Where do the problems come from? In some cases, they are problems I have read or heard elsewhere, recast in a different way. Some were suggested by a talk or conversation at a conference. Others were born of a particular experience—receiving a phone bill with an unexpected tax, flying between Australia and Canada, seeing a standard textbook problem, attending the annual booksale at University College of the University of Toronto. The problem about the folded sheet was a byproduct of trying to make up an elementary calculus question (p. 87). The problem about Humphrey the Horse came from a contest paper sent by a friend (p. 6). One of my favorites is the one about the ten digit number (p. 117), which was told to me by Anthony Gardiner and discussed on pages 24 to 34 in his very fine book, *Discovering Mathematics : The Art of Investigation* (Oxford: Oxford University Press, 1987). Also, many of the problems in this volume did not appear in the *Alumni Magazine*, but were passed on to me by L. J. Upton, of Mississauga, Ontario, who knows the amateur literature very well.

Mathematics, like music, can potentially reward almost everyone in some way. While music has developed a whole spectrum of devotees, from the serious professional to the casual listener, mathematics unfortunately has not succeeded in reaching its natural audience. Perhaps, unlike music, its position as a compulsory school subject has deterred people from discovering its more creative and recreational side. There is a vast literature accessible to the amateur mathematician, and, if this volume encourages the reader to seek it out and explore it, I shall be very pleased.

In the meantime, browse through this volume, pick up a problem and have a go. Almost all of the problems require no more than the mathematics of elementary school along with some basic reasoning and sensitivity to patterns; one or two will need a little high school mathematics. Have confidence in your own ability. If you are patient, you may find that you can solve more than you ever thought. Enjoy yourself!

I would like to thank my wife, Eileen, for her invaluable help and encouragement.

Chapter One

For Starters

A SCISSORS AND PAPER PUZZLE

Take a rectangular sheet of paper, and using a pair of scissors, reproduce the object pictured below. No use of glue or scotch tape is permitted.

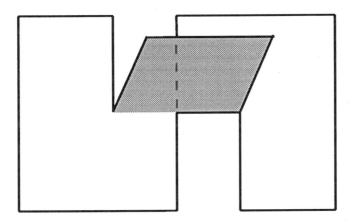

Success with this puzzle will give you a nice conversation piece for your coffee table.

Solution

Cut the sheet along the solid vertical lines as indicated below. Fold along the dotted line to make a vertical flap. Twist the shaded area down.

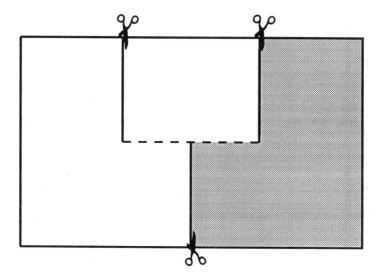

TWO CAKE-SLICING PROBLEMS

1. Suppose that you are given a fruitcake in the shape of an ordinary brick. How would you divide it among eight people with three straight cuts so that each person gets exactly the same amount of cake?

2. Given a square chocolate cake of uniform height that is iced on top and on its sides, divide it among nine people so that each person gets exactly the same amount of cake and icing.

Comments and Extensions

1. What is the maximum number of pieces that a cake can be divided into with a single cut? Yes, the answer is obviously two. If we are going to get up to eight pieces in three cuts, each piece will have to be divided into two with each new cut. You might see how this could be done by stacking the pieces after each cut, but can you solve the problem without moving any part of the cake at any time?

   This suggests the question of determining the maximum number of pieces (not necessarily with each having the same volume) that the cake can be divided into with four cuts; with five cuts; etc., without moving the pieces.

   Another problem applies not to a cake, but to a doughnut (or bagel, that is, the shape of a torus). What is the maximum number of pieces that can be obtained by cutting a doughnut with three straight slices of a knife?

2. You may have found a solution to the problem. But ask yourself: how simple is it? If you have not used all of the cake, then you are being unnecessarily stingy. If you cut the cake into a whole lot of little pieces and gave at least some of the nine more than a single piece, then you are being unduly messy. It is possible to solve this problem by using straight cuts and giving each person a single piece of cake. *Hint*: How would you solve the problem with a circular cake?

Solutions

1. The first two cuts are vertical, each joining the midpoints of opposite sides of the top. The third cut is horizontal and passes through the middle of the cake.

2. Here is a second-best solution. If there were eight people, the task is easy: slice from the centre to the four corners and the midpoints of the sides. We can adapt this for nine. Cut the ninth piece first from the middle of the cake as follows:

   - draw a square in the centre of the top of the cake whose area is equal to one-ninth the area of all of the icing;

   - draw a square in the centre of the bottom of the cake so that the solid portion obtained by making straight slices joining corresponding sides of the two squares is one-ninth of the volume of the cake.

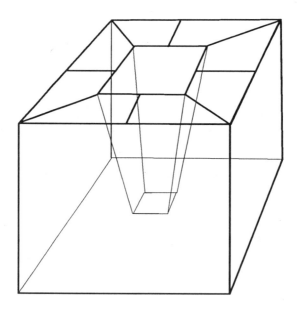

The ninth piece is the truncated pyramid so defined. The remaining eight pieces can be cut out by slicing from the centre.

This solution has a lot to commend it. However, it is dependent on the height of the cake. If the cake is *very* tall, the amount of icing on top may be less than one-ninth of the total amount and we should rather cut in from the sides of the cake. In addition, we may get too much cake by taking the ninth slice right to the bottom; we may have to take an inverted pyramid instead. But enough of these complications—there is a better way !

Moving around from any starting point, divide the perimeter of the top of the cake into nine equal parts. Make vertical cuts from the division points to the centre of the cake. Each person gets the same amount of icing from the top and from the sides; each person also gets the same volume of cake.

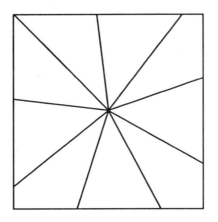

THREE PUZZLES WITH MATCHES

1. By moving just one match in the configuration
    below, produce a new figure similar to the original.

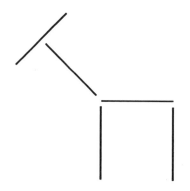

(The final figure need not be oriented the same way.)

2. This puzzle comes from the UK Schools Mathemati-
    cal Challenge, a multiple choice competition for
    students not beyond their ninth year in school.

    Humphrey the Horse at full stretch is hard to match.
    But that is what you have to do: move one match to
    make another horse just like (i.e., congruent to)
    Humphrey. Which match — A, B, C, D, E, — must
    you move?

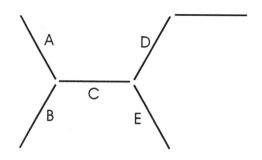

3. The matches in the configuration below surround five equal squares. By moving three of them, obtain a configuration for which all the matches surround exactly four equal squares.

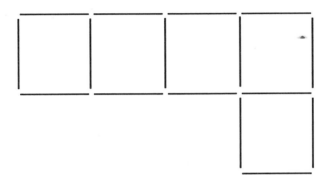

Discussion

1. & 2. One strategy to follow in problems like these is to see the figure as almost symmetrical. Can you find an axis of reflection, for example, for which the reflected figure is the same as the given figure except for one displaced match? Alternatively, one might look for a rotation of the figure about a point in the plane with the same property.

2. The obvious match to move, of course, is the one that is not labelled. Try removing each remaining match to see whether the figure that remains is symmetrical.

3. How many matches are there? These matches have to surround four squares; how many of them would be common sides of two squares?

Solutions

1. Move the right hand vertical match into a horizontal position.

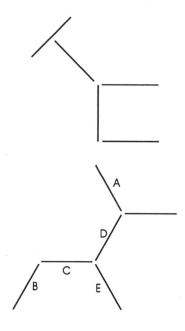

2. Consider a reflection in the axis that contains the match E. This reflection takes A to its new position at one end of D.

3. Either

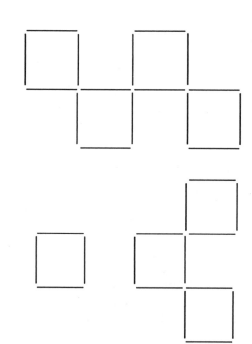

       or

FIVE RECTANGLES MAKE A SQUARE

F ind five rectangles, having sides with lengths cho-
sen from the numbers 1, 2, 3, 4, 5, 6, 7, 8, 9, 10 (each number used only once for a pair of opposite sides) which can be put together with no overlapping and no empty holes to form a square.

Suggestions

What are the possibilities for the side length of the square? One way to find this is to pair the set of 10 numbers and add together the five products of the pairs, getting a possible total area for the 5 rectangles. What is the largest value you can make this sum by suitably selecting the pair? The smallest?

For example, we could take the pairs (1, 4), (2, 6), (3, 5), (7, 9), (8, 10) for opposite sides of the rectangles, which yields $4 + 12 + 15 + 63 + 80 = 174$, which is not a square number. Clearly, we cannot make a square from rectangles with these five pairs of dimensions? Can we make a larger sum, say 196? A smaller, say 169?

Solutions

Let us first find an upper bound for the total area of the rectangles. If 10 and 9 are not paired, we would have pairs (10, a) and (9, b) for some numbers a and b. Since

$$10 \times 9 + ab - (10a + 9b) = (10 - b)(9 - a),$$

which is positive, $10 \times 9 + ab$ exceeds $10a + 9b$. Thus, if 10 and 9 are not paired, we can make the total area of the rectangles larger by pairing 10 and 9 and pairing their previous mates.

We can similarly argue that we should pair 8 and 7, then 6 and 5, etc. to make the total area as large as possible. Thus, the total area of all five rectangles cannot exceed

$$10 \times 9 + 8 \times 7 + 6 \times 5 + 4 \times 3 + 2 \times 1 = 190.$$

A similar argument establishes that the total area of all the rectangles must be at least

$$10 \times 1 + 9 \times 2 + 8 \times 3 + 7 \times 4 + 6 \times 5 = 110.$$

Therefore, if the five rectangles can be put together to form a square, the square must have one of the areas 121, 144, or 169.

Here are four possible solutions to the problem:

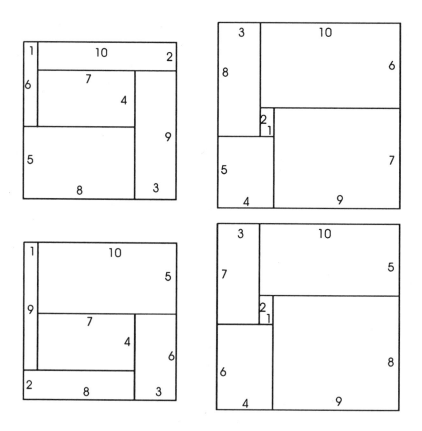

TWO PROBLEMS OF TRANSFERENCE

1. Three baskets contain, respectively, 6, 7, and 11 marbles. By moving marbles from one basket to another, transferring into a basket only as many marbles as are already there, arrange it so that there are eight marbles in each basket.

2. Three unmarked jars have capacities of 19, 13, and 7 litres respectively. The first is empty while the other two are full of fluid. How can you measure out 10 litres using no other vessel, merely by transferring fluid from one container to another?

Hints

While a certain amount of inspired trial and error might get you to the solution, there are a couple of systematic approaches that might be taken. One is to make a "tree" of possibilities. Denote by (a, b, c) the state in which the three containers have respectively a units, b units, and c units. We can indicate all possible states obtainable from (a, b, c) by lines and continue on. For example, in the first problem, we can make the tree:

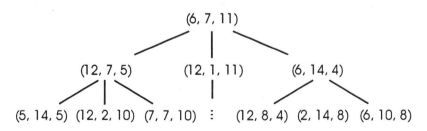

Another strategy is to work backwards from the end result.

Solutions

1. If you got a solution in 6 moves, you did pretty well.

Here's one:

(6, 7, 11), (12, 7, 5), (12, 2, 10), (10, 4, 10),
(6, 8, 10), (12, 8, 4), (8, 8, 8).

But then, you might have done better with a 4-move solution:

(6, 7, 11), (12, 1, 11), (12, 2, 10),
(12, 4, 8), (8, 8, 8).

Did you realize that you can actually get by with three moves?

(6, 7, 11), (6, 14, 4), (12, 8, 4), (8, 8, 8).

2. The following solution takes 15 moves:

(0, 13, 7), (7, 13, 0), (19, 1, 0), (12, 1, 7), (12, 8, 0),
(5, 8, 7), (5, 13, 2), (18, 0, 2), (18, 2, 0), (11, 2, 7),
(11, 9, 0), (4, 9, 7), (4, 13, 3), (17, 0, 3), (17, 3, 0),
(10, 3, 7).

If we wish, we could take an extra move to end up with (10, 10, 0).

MAKING A PATIO TO A "T"

At a gardening sale, Sam picked up 15 T-shaped landscaping slabs as pictured below:

Since the area of each slab is 4 square feet, Sam decided that he would use them to construct a 60 square-foot rectangular patio. How should he arrange them?

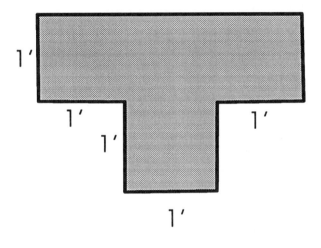

Solution

The task is, alas, impossible. Let us picture the proposed rectangular pattern as being ruled into 60 squares, each having sides of one foot. Suppose that these squares are coloured white and black like a checkerboard. Then there must be 30 squares of each colour.

Now let us try to place the T-shaped slabs. Each slab must cover either three white and one black square, or three black and one white square. In either case, the slab must cover an odd number of squares of each colour. Therefore, fifteen slabs must cover an odd number of squares of each colour, so they certainly could not cover 30 squares of each colour.

NOT A KNOT

The mathematical field of knot theory arose from a 19th-century attempt to account for the properties of atoms by knots in the ether. While it is now pursued for its intrinsic interest, the theory does have a role in fluid flow, DNA research, and theoretical physics.

Readers who were Girl Guides and Scouts will undoubtedly recall having tied a reef knot (square knot) with a single piece of rope (Diagrams A and B). There is an unusual way of "cancelling" a reef knot. To make it work properly, you should have a rope or a fairly robust piece of twine. Take the end marked X and pass it under, then over, its neighbouring strand (Diagram C). Next pass the end X underneath itself and up through the middle of the knot (Diagram D). Finally, pull X and Y away from each other and the knot should come out. Try it and see. In the jargon of knot theory, Diagram D actually represents an unknot.

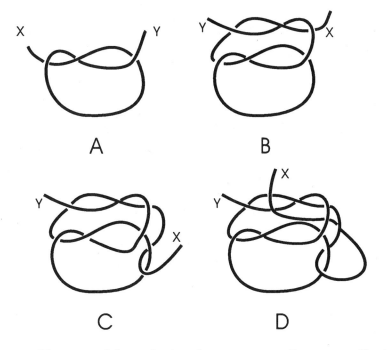

Your problem is to draw some diagrams that will illustrate that Diagram D is in effect not a knot at all.

Comments

This problem drew responses from five people when it was originally posed. One of the five also provided a super-sonnet (a sonnet with an extra line). I give you this knot-sensical creation of Ellen Carlisle of Toronto:

> To be or not to be a knot -
> A knotty puzzle is it not?
> Unknotted knots, not knotted knots
> Are unknots to the Knights of Knots.
> Why is an unknot not a knot?
> A knitted knot you'll see cannot
> Unknot itself; but not unknots.
> Knots, assume correct position !

But naughty unknots never listen.
So unknots lose their knottiness
When they engage in naughtiness,
Resulting in the cancellation
Of their knotty reputation.
Or perhaps a confirmation
Of their naughty occupation.

Solution

The top diagram is easily obtained from Diagram D by joining the ends *X* and *Y* together.

Now draw *XY* back through the two positions marked with the arrowheads, and get the situation in the bottom diagram.

Now draw *XY* back through the two positions indicated in the second diagram, and you will see that the knot comes out very nicely. [AF]*

---

*Initials appearing in square brackets are those of contributors of solutions and suggestions used in the text. For names see the list of contributors in the Appendix.

PRINTER'S PUZZLE

The photocopier used by Angela for her monthly newsletter feeds the paper in from a hopper on the right, the image is printed on the upper side, and the final copy is delivered face up into a tray on the left. The original is placed face down on the glass with the top of the page away from the user; however, the copy in the tray has the top toward the user (so it is upside-down). For two-sided printing, paper is taken from the tray on the left and placed plain side up in the hopper on the right.

Angela's newsletter consists of four pages printed on a single sheet in booklet format. Pages 1 and 4, typed on separate sheets, are duplicated together on one side (reduced); pages 2 and 3 are similarly duplicated on the other. Explain how Angela places the originals and feeds in the paper for copying.

For a greater challenge, consider how 16 pages might be printed on two sides of a single large sheet, which then can be folded and certain edges cut to produce a book. This is the sort of problem a publisher has to solve in producing a "signature" for a book.

Solution

For the photocopying problem, we want to arrange that page 1 is on the back of page 2, and that page 3 is on the back of page 4. The two sides of the sheet of paper are to be

$$\boxed{4 \mid 1} \quad \text{and} \quad \boxed{2 \mid 3}$$

Place the original of the 4-1 side face down on the glass with page 1 away from the user. Take the printed pages from the tray and, keeping the same edge towards the user, refeed them blank side up into the hopper. For printing the 2-3 side, place the original face down on the glass with page 2 away from the user.

There are other ways of proceeding, for example, by positioning pages 4 and 3 away from the user on the glass, or by putting 1 and 3 away from the user on the glass and giving the copy pages a 180° rotation before feeding them into the hopper for printing the second side.

For the printer's signature of 16 pages, we want to arrange matters so that there are 8 pages on each side of the sheet, arranged in such a way that the sheet can be folded and slit leaving the 16 pages in order. In helping us think how this might be done, we should note that pages x and $17-x$ are adjacent, and that page $2y$ should be behind page $2y-1$. Three possible ways of organizing the pages are:

I.

5	12	$\overline{9}$	8		7	10	11	$\overline{6}$
4	13	16	1		2	15	14	3

(top rows shown inverted)

II.

13	4	5	12		11	$\overline{9}$	3	14
16	1	8	9		10	7	2	15

(top rows shown inverted)

III.

1	16	$\overline{9}$	8		7	10	15	2
4	13	12	5		6	11	14	3

(top rows shown inverted)

Chapter Two

Playing with Numbers

A LITTLE BIT OF ARITHMETIC

1. Dick burst in on his parents to get help with the following homework problem:

   A man standing in line for a movie observes that $\frac{5}{6}$ of the line is in front of him and $\frac{1}{7}$ of the line is behind him. How many are there in the line altogether?

   "Well," began Father, tamping his pipe, "I would let x be the number of ..."

   "Oh nonsense," shot in Mother, "the answer is clearly 42 !" Dick checked with paper and pencil. He turned to his mother with awe. "How did you know?"

   "If there is an answer to the problem, it has to be the lowest common denominator of the two fractions."

   Dick looked puzzled. "I don't think the teacher wants us to do it that way."

   "Well," suggested Mother, "you can't just pull the answer out of the blue. But if you can give a sound argument to narrow the answer down to one possibility, and if you verify whether it works I can't see that there would be much objection. Now run along and see what you can do."

What was the reasoning that Mother had in mind? Can her method be applied if $\frac{5}{6}$ and $\frac{1}{7}$ are replaced by other fractions?

2. The school secretary looked most severely at Mr. Chalk as he picked up his mail. "Mr. Chalk, you have not given me your class number." "Oh yes, I have it right here: 15 231 236 267 520. It took me quiet a while to work this out, multiplying together all the ages of the students in my class. But, remind me why you need it." "The Board of Education wants to store it in its computer. Apparently, from this class number, it is possible to tell how many students there are in your class and what their ages are."

"That seems strange. A class with eight 12-year-olds would give me the same product of ages as one with three 16-year-olds and four 18-year-olds. How would I distinguish between them?"

"You forget this is a high school. We have only teenagers here. Nobody is under 13 or over 19. In this situation, the class number creates no ambiguity."

Is the secretary correct? What are the ages of the students in Mr. Chalk's class? [JG-McL]

3. In 1988, the government of the Province of Ontario imposed a tax on telephone bills. The Federal Finance Minister glowered at the phone bill on his desk and muttered to his aide, "Péntek, those provincial rascals have imposed their sales tax not only on the bill, but on our tax as well. Our rate is 10 per

cent and their tax is 8 per cent. On a $100 account we levy a tax of $10 and they charge 8 per cent of $100 plus $10 for a total tax of 8.80."

"Highly irregular, sir."

"You bet it is ! Why aren't we taxing *their* tax?"

"But then they would be obliged to tax our extra tax, whereupon we would have more of their tax to tax. But then, ... it would go back and forth for ever. Who knows what it might amount to !" The Minister rubbed his hands. "Yes ! I am sure you chaps can come up with a formula." The FFM mused, "If we were to tax the base amount and their tax at 10 per cent, and they were to tax the base and our tax at 8 per cent, how much revenue can the federal and provincial governments expect to get from a phone bill of $100?"

Péntek scurried away. Help him find an answer. Also, determine the revenue if each government taxes its own tax as well.

Péntek suggested this to the FFM, but it was felt that this was going a little far in an election year.

Comments

1. In the movie theatre lineup, the fractions given of the length of the line represent numbers of people and therefore must be integers. What does this tell you about the relationship between the denominators and the total number? How would you characterize the numbers for which $\frac{5}{6}$ and $\frac{1}{7}$ of them are whole numbers?

2. Are there any ages for which one can tell without ambiguity how many students there are?

3. One way to tackle this problem is to make a table with two headings:
   federal tax and *provincial tax.*

   In the first line, enter the taxes on the base amount. In each subsequent line, enter the taxes due from the previous line. The amounts, of course, will get smaller as you go down. Alternatively, you can treat it as a problem in algebra; let F and P be the total federal and provincial taxes due, and set up the equations.

Solutions

1. Here is how Mother's approach can be justified. Since $\frac{5}{6}$ of the total number must be an integer, the total number must be divisible by 6. Similarly, it also must be divisible by 7. Therefore, the number of people in the theatre line must be a multiple of $42 = 6 \times 7$. We next argue that it cannot be any higher multiple than 42 itself.

   The fraction of people left in the theatre line beyond the $\frac{5}{6}$ and $\frac{1}{7}$ has a denominator equal to 42. If we multiply this by a higher multiple of 42, say 84, or 126, then surely the number of people represented will be strictly greater than 1. Thus, if there is an answer, 42 is the only possibility.

   We can now check that 42 works. There are 35 people in front of the man in the problem, and 6 behind; this accounts for 41 people. Including the man gives 42.

2. Factored into primes, the calculated class number,

   15 231 236 267 520,

   is equal to $2^9 \, 3^5 \, 5^1 \, 7^3 \, 13^1 \, 17^2 \, 19^1$. Given the age-range of the students, each one of the primes 5, 7,13, 17 and 19 can be divided into only one age of student. Thus, we see right away that there must be three 14-year-olds, one 15-year-old, one 13-year-old, two 17-year-olds and one 19-year-old. Since $15 = 3 \times 5$, the 15-year-old accounts for one of the factors 3; the other four factors 3 must be due to the 18-year-olds. Since $18 = 2 \times 3 \times 3$, there must be two 18-year-olds for two 2's, leaving four 2's to account for the one 16-year-old.

   In summary, we have:

   $$15 \ 231 \ 236 \ 267 \ 520 = 13^1 14^3 15^1 16^1 17^2 18^2 19^1,$$

   and the exponents for each possible age indicates the number of students of that age.

3. Taking the base amount as $100 with a federal tax rate of 10% and a provincial tax rate of 8%, let F be the total federal tax payable and P be the total provincial tax payable. Assuming that each government taxes the base amount and the tax of the other government (but not its own), we see that F and P satisfy the system of equations:

   $$(100 + P)(0.10) = F$$

   $$(100 + F)(0.08) = P.$$

   These simplify to $100 + P = 10F$ and $800 + 8F = 100P$ whence $992F = 10{,}800$ and $992P = 8800$. Thus, F = 10.89 and P = 8.87, so that the federal tax is $10.89 and the provincial tax is $8.87. The taxes total $19.76, almost twenty per cent.

We could also view this situation in terms of a number of "rounds." In the first round, each government taxes only the base amount. In each subsequent round, each government taxes the tax of the other government found in the previous round. This table gives the amounts:

Round number	Federal Tax	Provincial Tax
1	$10.0000	$8.0000
2	0.8000	0.8000
3	0.0800	0.0640
4	0.0064	0.0064
5	0.0006	0.0005
6	0.0001	0.0000

By this stage, the added taxes have become negligible. Adding and rounding to the nearest cent yields taxes of $10.89 and $8.87. Observe that the taxes accruing to the two governments on the even rounds are equal.

The assumption that each government rapaciously taxes its own as well as the other's tax leads to the equations

$$(100 + P + F)(0.10) = F$$

$$(100 + P + F)(0.08) = P.$$

In this case, the federal tax is $12.20 and the provincial tax is $9.76. It is interesting to observe that, when the provincial government taxes the federal tax but the federal government does not tax the provincial tax, then the tax apportionment is:

Federal: $10.00
Provincial: $8.80
Total: $18.80.

If only the federal government does the extra taxing, the apportionment is:

Federal: $10.80
Provincial: $8.00
Total: $18.80.

The total tax is the same either way. This suggests that if we want to arrange for both governments to get an extra tax bite without increasing the total, a fair way of sharing the loot might be:

Federal: $10.40
Provincial: $8.40
Total: 18.80.

THE NINE-DIGIT NUMBER

A certain 9-digit number uses each of the non zero digits 1 2, ..., 9 exactly once. As can be seen by casting out nines (see the Appendix), this number is necessarily divisible by 9.

However:

the 8-digit number obtained by removing the last digit (the units digit) is evenly divisible by 8;

the 7-digit number obtained by removing the last two digits is evenly divisible by 7;

the 6-digit number obtained by removing the last three digits is evenly divisible by 6;

*   *   *

the 2-digit number obtained by removing the last seven digits is evenly divisible by 2.

Determine the number.

Hints

What can be said about the positions of the odd digits? The even digits? What about the fifth digit? Information on the first three digits can be provided by an extension of the rule of casting out nines: a number is divisible by 3 if and only if the sum of its digits is divisible by 3.

Solution

We can narrow down the possibilities by a sequence of steps:

(1) Since 0 is not present, the fifth digit must be 5.

(2) Since multiples of 2, 4, 6, 8 are even, the even digits must be in the even positions, and therefore the odd digits are in the odd positions.

(3) Since any multiple of 200 is automatically divisible by 8, the seventh and eighth digits form a number divisible by 8, and so the digits must be 16, 32, 72, 96.

(4) Since any multiple of 100 is automatically divisible by 4, the third and fourth digits form a number divisible by 4. Since the third digit is odd, the fourth digit is either 2 or 6.

(5) Since the first three digits make up a number divisible by 3, as do the first six digits, the fourth, fifth, and sixth digits make up a number divisible by

3. Hence, both the sum of the first three digits and the sum of the second three digits is divisible by 3.

(6) From the foregoing assertions, we conclude that the number has one of the following four forms:

$$\_8\_65432\_$$
$$\_8\_65472\_$$
$$\_4\_25816\_$$
$$\_4\_25896\_$$

(7) At this point, some trial and error, and the fact that the first seven digits give a number divisible by 7, reduces the number of possibilities to one: 381654729.

FINDING CUBE ROOTS IN YOUR HEAD

Quickly, in your head, given that it is a whole number, find the cube root of 317 214 568. (It is the product of three equal integers, that you must find.) Don't worry—this is not a big deal ! The answer has only three digits, and you have lots of information. If the last digit is 0, 1, 4, 5, 6, or 9, the cube root has the same last digit. The digits 2 and 8 get swapped between cube and cube root, as do 3 and 7. Now you know the last digit of the root is 2.

Now, partition the digits in groups of three, working from the right. Look at the left block (317). At this stage, you should have memorized the first ten cubes (0, 1, 8, 27, 64, 125, 216, 343, 512, 729). Since 317 lies between the cubes of 6 and 7, you seek a number whose first digit is 6.

The middle digit can be found by casting out 3s. The sum of the digits in the given number is 37, and the remainder upon division by 3 is 1. Summing the digits of the cube root (6 and 2 and the unknown digit) should give the same result. Thus, the desired cube root is either 622, 652, or 682. Since 317 is reasonably close to 343, much closer than to 216, the answer is probably 682. This turns out to be correct. Half an hour's practice with a friend giving you cubes from a calculator will make you quite expert—and perhaps you will devise some short cuts.

Here are some examples for you to try. Find the cube roots of 571 787; 2 000 376; 45 499 293; 167 284 151; 1 473 760 072. (Answers at the bottom of page 30.)

Some Comments

The method suggested rather critically depends on the number being an exact cube. One suggestion is to memorize a few logarithms (to base 10): log 1 = 0, log 10 = 1, log 2 = 0.30, log 3 = 0.48 (a little less than $\frac{1}{2}$, since 3 is a smidgen less than the square root of 10). From these, you can get quickly the logarithms of 4 (0.60), 8 (0.90), 6 (0.78).

Now 317 214 568 is $3.17\ldots\times10^{8}$. And 3.17 is so close to the square root of 10 that its logarithm should be just about 0.50. So the logarithm of our number is 8.50. One third of 8.50 is 2.83, so the number we're looking for is 100 times 10 raised to the 0.83.

Over a small range you can get away with linear interpolation. We know that log 6 is 0.78 and log 8 is 0.90. So 0.83 should be the logarithm of $6 + \frac{5}{12} \times (8 - 6)$, which is a bit more than 6.8. The cube root of 317 214 568 is therefore a bit more than 680.

If you know that you were given an exact cube, then you can be sure that the answer is 682. [J.S.M]

THE DEFECTIVE CALCULATOR

A certain pocket calculator can add, subtract, and take reciprocals of numbers. However, the times (multiplication) button is defective, so that it cannot multiply. Of course, given two whole numbers, we could perform a multiplication by simply adding one of them to itself an appropriate number of times. However, this is a tedious task. How else is it possible to find the product of two given numbers by using only the three available operations?

Solution

There are many ways to solve the problem. Here is one:

(1) First, we need to know how to square a number, x. If $x = 1$, this is easy, so let us assume that $x \neq 1$. Observe that

$$\frac{1}{x} + \frac{1}{1-x} = \frac{1}{x - x^2}.$$

Subtract x from 1, and take the reciprocal of $1 - x$. Add to this the reciprocal of x, and then take the reciprocal of the sum to obtain $x - x^2$. Finally perform a subtraction to get $x^2 = x - (x - x^2)$.

(2) Now that we know how to square, we can get four times the product of two numbers by using $4xy = (x + y)^2 - (x - y)^2$.

(3) Finally, we need to divide by 2 twice. This can be done by adding the reciprocal of the number to be halved to itself, and taking the reciprocal of the result, since $\dfrac{2}{x} = \dfrac{1}{x} + \dfrac{1}{x}$.

Alternatively, to find one quarter of the square of a given number, we can exploit the identity

$$a + \left[\frac{1}{a-4} - \frac{1}{a} \right]^{-1} = \frac{a^2}{4}.$$

Then use $xy = \dfrac{(x+y)^2}{4} - \dfrac{(x-y)^2}{4}$ to find the desired product. [D.C.B.]

Answers to problems on page 28: 83; 126; 357; 551; 1138

PATTERNS

There is a strong experimental side to mathematics. It is the science of structure, and the first step is to discover what structure is there by simple observation. This requires sensitivity to detail and imagination in guessing possible patterns. Only then can we formulate conjectures and prove theorems.

Take, for instance, the Fibonacci sequence, which has been fascinating people for about 800 years:

$$1, 1, 2, 3, 5, 8, 13, 21, 34, 55, \ldots$$

The three dots mean that we continue the sequence indefinitely.

What terms should follow 55? Let us look for regularities. Each term from the third is the sum of its two predecessors:

$$2 = 1 + 1$$
$$3 = 2 + 1$$
$$5 = 3 + 2$$
$$8 = 5 + 3.$$

Assuming this pattern holds, we can continue:

$$89, 144, 233, \ldots .$$

The Fibonacci sequence is rich in interesting relationships. The square of each term differs from the product of its immediate neighbours by 1 or -1:

$$3^2 - 5 \times 2 = -1$$
$$8^2 - 5 \times 13 = -1$$
$$13^2 - 8 \times 21 = 1.$$

Adding the squares of two adjacent terms gives a term later in the sequence:

$$3^2 + 5^2 = 9 + 25 = 34$$

$$5^2 + 8^2 = 25 + 64 = 89.$$

The reader might discover other patterns. The fecundity of the Fibonacci sequence has inspired a journal, *The Fibonacci Quarterly*, devoted to its properties.

Before we get to some problems, let me introduce you to a notation which is convenient for expressing our results. Terms of a sequence can be indicated by a subscripted notation in which a "tag" attached to a letter indicates how far along we are in the sequence. For example, the Fibonacci sequence is often indicated by the letter F (naturally). We write

$$F_1 = 1 \; ; \; F_2 = 1 \; ; \; F_3 = 2 \; ; \; F_4 = 3 \; ; \; F_5 = 5 \; ; \; F_6 = 8 \; .$$

The general, or the nth-term, is denoted by F_n.

The patterns described above can be written

$$F_n^2 - F_{n-1}F_{n+1} = (-1)^{n-1} \quad \text{for } n = 2, 3, 4, 5, \ldots$$
$$F_n^2 + F_{n+1}^2 = F_{2n+1} \quad \text{for } n = 1, 2, 3, 4, \ldots$$

Problems

1. Here are the beginnings of a few sequences. Decide what the next few terms should be, and discover as many interesting patterns as you can:

   (a) 1, 3, 4, 7, 11, 18, 29, 47, ...

   (b) 1, 9, 36, 100, 225, 441, ...

   (c) 3, 5, 13, 85, 3613, ...

   (d) 1, 1, 2, 5, 14, 42, 132, 429, ...

2. Consider the following sequence:

$$11, 34, 17, 52, 26, 13, 40, 20, \dots$$

What is the rule that determines each term from its predecessor?

Continue this sequence further and see what happens.

The first term in the sequence can be anything you like. Experiment with sequences having the same rule of formation and different first entries.

3. Sometimes pattern recognition will lead us to new solutions of equations when some solutions are already known. *Diophantine equations* are those for which we want solutions in which the variables take integer, or whole number, values. The most well known Diophantine equation is without doubt the Pythagorean equation:

$$x^2 + y^2 = z^2.$$

(a) There are many solutions for which $z = y + 1$. Some of these are $(x, y, z) = (3, 4, 5)$, $(5, 12, 13)$, $(7, 24, 25)$. Find others.

(b) There are many solutions for which $y = x + 1$. Some of these are $(x, y, z) = (3, 4, 5)$; $(20, 21, 29)$; $(119, 120, 169)$. Find others.

If you know a little algebra, you can avoid multiplying out the squares when you check your results. Use factorization. For example:

$$29^2 - 21^2 = (29 - 21)(29 + 21)$$
$$= 8 \times 50 = 16 \times 25 = 4^2 5^2 = 20^2.$$

4. There is a cubic version of the Pythagorean equation:

$$x^3 + y^3 + z^3 = w^3.$$

Here are the first few cases in families of solutions this equation; see if you can find other solutions.

(a) $9^3 + 15^3 + 12^3 = 18^3$
   $28^3 + 53^3 + 75^3 = 84^3$
   $65^3 + 127^3 + 248^3 = 260^3$

(b) $3^3 + 4^3 + 5^3 = 6^3$
   $3^3 + 10^3 + 18^3 = 19^3$
   $12^3 + 19^3 + 53^3 = 54^3$
   $12^3 + 31^3 + 102^3 = 103^3$
   $27^3 + 46^3 + 197^3 = 198^3$
   $27^3 + 64^3 + 306^3 = 307^3$

5. What is the rule that governs the following arrangement of digits?

$$8\ 5\ 4\ 9\ 1\ 7\ 6\ 3\ 2\ 0$$

6. What should the next two terms in the following sequence be?

0, 1, 10, 2, 100, 11, 1000, 3, 20, 101, ___, ___

Hints

1. The first thing to check is whether all the numbers in the sequence have any characteristic property which makes them all of the same type. If this fails, one might suppose that each term in the sequence arises from its predecessor(s) in some way.

2. Which terms in the sequence are followed by a bigger term, and which by a smaller?

3. (a) Look at the square of x and compare with $y + z$.

(b) Go "back" one step and include the degenerate solution $(x, y, z) = (0, 1, 1)$. Look for patterns in the sequence of z-terms:

$$1, 5, 29, 169.$$

Also, look at the odd number of the x and y pair in each solution:

$$1, 3, 21, 119.$$

Factor these numbers.

4. (a) Subtract 1 from x in each solution; what do you notice? Also examine $w - z$ and $\dfrac{w}{x}$.

(b) Look at $\dfrac{x}{3}$; $y - x$; the ratio of the largest even number to x.

5. The French version is to find the rule for

$$5\ 2\ 8\ 9\ 4\ 7\ 6\ 3\ 1\ 0.$$

6. This is a representation of a well-known sequence. When is the last digit nonzero? The second last digit?

Solutions and Comments

1. There is no single correct answer to these sequences, as different people will detect different patterns which may or may not yield the same answer. There may be differences of opinion as to how the sequences might be continued; in this situation, one applies an aesthetic criterion—which "theory" of the sequence is most simple or attractive?

(a) This one is a relative of the Fibonacci sequence (called the Lucas sequence). Each term is the sum of its two predecessors. Other patterns: the square of each term differs from the product of its neighbours by ±5; the sum of the squares of two adjacent terms is 5 times a term in the Fibonacci sequence. The following relations connect terms of the Fibonacci and Lucas sequences (W.K.):

$$4 = 3 \times 1 + 1 \times 1$$
$$7 = 4 \times 1 + 3 \times 1$$
$$11 = 4 \times 2 + 3 \times 1$$
$$18 = 7 \times 2 + 4 \times 1$$
$$29 = 7 \times 3 + 4 \times 2$$
$$47 = 11 \times 3 + 7 \times 2$$

(b) All the terms in the sequence are squares, and the difference between two adjacent terms are cubes:

$$9 = (1 + 2)^2 = 1^3 + 2^3$$
$$36 = (1 + 2 + 3)^2 = 1^3 + 2^3 + 3^3,$$

and so on.

(c) Each term is found by squaring its predecessor, adding 1 and then dividing by 2. There are other ways of generating the sequence:

- Subtract 1 from each term and take half to get the sequence:

  1, 2, 6 (=2×3), 42 (=6×7), 1806 (=42×43).

- Note that half the differences between successive terms is a perfect square:

  1, 4 (=2^2), 36(=6^2), 1764(=42^2).

(d) This is a tougher nut to crack. To describe one law of formation conveniently, let us introduce the idea of a vector, or ordered list of numbers, $(a, b, c, ..., k)$. The dot product

$$(a, b, c, ..., k) \cdot (p, q, r, ..., z)$$

of two vectors of the same length is found by taking the products of corresponding entries and adding them:

$$ap + bq + cr + ... + kz.$$

Applying this notation to the sequence at hand yields:

$$2 = (1, 1) \cdot (1, 1);$$

$$5 = (1, 1, 2) \cdot (2, 1, 1);$$

$$14 = (1, 1, 2, 5) \cdot (5, 2, 1, 1); ...$$

Continuing on, we see that the term following 429 should be 1430.

Another way of building the sequence is to note that each term is the product of the previous term and a fraction whose numerator increases by 4 each time while its denominator increases by 1. Thus:

$$1 = 1 \times \frac{2}{2}; 2 = 1 \times \frac{6}{3}; 5 = 2 \times \frac{10}{4}; 14 = 5 \times \frac{14}{5}; ...$$

Note that, for $n = 3, 4, ...,$

$$1 + 24 \left[\frac{x_{n-1}x_{n-2}}{x_n x_{n-2} - x_{n-1}^2} \right]$$

is always a perfect square (W.K.); in fact, it is equal to $(2n - 1)^2$. For example:

$$1 + 24 \left[\frac{5 \times 14}{42 \times 5 - 14^2} \right] = 1 + 24 \times 5 = 11^2$$

This sequence is quite famous. Euler noted that the nth term is the number of ways in which a fixed regular polygon with $n+1$ vertices can be divided into triangles by diagonals which do not intersect. Here are the possibilities for a pentagon ($n = 4$):

Catalan (after whom the sequence is named) identifies the nth term as the number of ways in which a chain of n letters in a fixed order can be provided with $n-1$ pairs of parentheses in such a way that there are two terms between a corresponding pair. For example, when $n = 4$, we have

$(((ab)c)d)$, $((ab)(cd))$, $((a(bc))d)$, $(a((bc)d))$, $(a(b(cd)))$.

The mathematical formula for the nth Catalan number is

$$\frac{1}{n} \binom{2(n-1)}{(n-1)}$$

where the symbol $\binom{a}{b}$ stands for a fraction whose numerator is the product of all integers from $b + 1$ up to a and whose denominator is the product of all integers from 1 up to b :

$$\binom{a}{b} = \frac{(b+1)(b+2) \cdots (a-1)a}{1 \cdot 2 \cdots (b-1)b}$$

Here are a couple more sequences for you to chew on:

$$1, 3, 8, 21, 55, 144, \ldots$$

$$1, 2, 3, 7, 11, 21, 25, 35, 39, 49, 65, \ldots$$

(A.W.W.)

2. If a given term, say n, is odd, then the subsequent term is $3n + 1$; if it is even, the subsequent term is half. In practice, it has been found that, if one continues the sequence long enough (quite long if you start with some numbers, like 27), the sequence eventually resolves into the cycle 1,4, 2, 1. It is an open problem whether this happens regardless of the choice of first number.

3. (a) The solutions are all of the form

$$(x, y, z) = (2k+1, 2k(k+1), 2k(k+1) + 1)$$

where k is an integer. The key is to note that $y + z = x^2$.

(b) Two more are:

(696, 697, 985),   (4059, 4060, 5741).

Note that $29 = 5 \times 6 - 1$
$169 = 29 \times 6 - 5$
$985 = 169 \times 6 - 29$
$5741 = 985 \times 6 - 169.$

4. (a) (x, y, z, w)
$$= (q^3 + 1, 2q^3 - 1, q(q^3 - 2), q(q^3 + 1))$$

(b) (x, y, z, w)
$$= (3k^2, 6k^2 - 3k + 1, 3k(3k^2 - 2k + 1) - 1, z + 1)$$

$$(x, y, z, w)$$
$$= (3k^2, 6k^2+3k+1, 3k(3k^2+2k+1), z+1)$$

5. If each digit is written out as an English word, the digits are in alphabetical order.

6. 10000, 12.

   The entries code the prime power factorization of the consecutive numbers 1, 2, 3, ... The last digit gives the highest exponent of a power of 2 dividing the number, the second last digit the highest exponent of a power of 3 dividing the number, the third last digit the highest exponent of a power of 5, and so on.

A TRIO OF SEQUENCES

How does a mathematician decide what he wants to prove? Sometimes, results are suggested by patterns; a considerable amount of mathematical activity consists of experimentation to see if there are facts that may indicate a broader pattern. Number theory is a good area for this.

In the puzzle that follows, there is no one answer and you are limited only by your imagination. You will be given three sequences of numbers. There are relationships among their entries. Identify as many of these as you can.

Entry number	Soprano	Contralto I	Contralto II
0	0	0	1
1	1	1	1
2	6	2	3
3	35	5	7
4	204	12	17
5	1189	29	41
6	6930	70	99
7	40391	169	239
8	235416	408	577
9	1372105	985	1393

These sequences continue indefinitely. Can you supply the next few terms?

Getting Started

Look at the corresponding entries in the three sequences; do you see a simple way in which the soprano term is related to the contralto terms? Look at the sum and the difference of two consecutive soprano terms, and compare the results with the terms in the contralto sequences. Double the terms in the soprano sequence and see if the numbers appear elsewhere.

There is a systematic way in which the entries in each of the sequences is obtained from their predecessors. See if you can find it, and try to extend the sequences.

A pythagorean triple (a, b, c) is a set of three integers for which $a^2 + b^2 = c^2$. Two examples of pythagorean triples are $(3, 4, 5)$ and $(20, 21, 29)$. In both of these, the smaller two entries differ by 1. Can you use the sequences and patterns arising from them to generate other pythagorean triples whose smallest two numbers differ by 1?

Describing The Patterns

As we did earlier, we can describe patterns using a subscripted notation. We will use the letter x to refer to the soprano sequence, y for contralto I, and z for contralto II. For the individual terms of the sequence, we will make use of a subscript or index as a tag on the symbol for the sequence. It is convenient to start our figures with the number 0. Thus, x_3 will refer to entry number 3 in the soprano sequence: $x_3 = 35$.

We have:

$$x_0=0 \qquad y_0=0 \qquad z_0=1$$
$$x_1=1 \qquad y_1=1 \qquad z_1=1$$
$$\cdots \qquad\qquad \cdots \qquad\qquad \cdots$$
$$x_8=235416 \quad y_8=408 \quad z_8=577$$
$$\cdots \qquad\qquad \cdots \qquad\qquad \cdots$$

Let us look at a couple of examples to see how we can make use of this symbolism.

Doubling the terms of the soprano sequence, we find that

$2 \times 0 = 0$ — the zeroth entry of contralto I

$2 \times 1 = 2$ — the second entry of contralto I

$2 \times 6 = 12$ — the fourth entry of contralto I

$2 \times 35 = 70$ — the sixth entry of contralto I

Using the subscripted notation, we can rewrite these as:

$$2x_0 = y_0$$
$$2x_1 = y_2$$
$$2x_2 = y_4$$
$$2x_3 = y_6$$

We notice that the subscript for the x-sequence is half that for the y-sequence. Continuing the pattern, we might well suspect that

$$2x_{27} = y_{54}$$

Use the letter n to represent a general entry number. Then it would appear that $2x_n = y_{2n}$ (i.e., double the nth entry of the x-sequence gives the $(2n)$th entry of the y-sequence).

For our second example, consider the terms of the contralto I sequence. We observe that

$$2 = 2 \times 1 + 0 \qquad y_2 = 2y_1 + y_0$$
$$5 = 2 \times 2 + 1 \qquad y_3 = 2y_2 + y_1$$
$$12 = 2 \times 5 + 2 \qquad y_4 = 2y_3 + y_2$$
$$29 = 2 \times 12 + 5 \qquad y_5 = 2y_4 + y_3$$

Evidently, each term depends on its two predecessors is the same way. We can express this by the statement:

$$y_{n+1} = 2y_n + y_{n-1} \quad \text{for } n \geq 1.$$

Setting $n = 3$, for example, gives $n + 1 = 4$ and $n - 1 = 2$, as in one of the equations in our list above.

Before turning the page, see how many other patterns you can find.

Some Patterns

1. $x_{n+1} = 6x_n - x_{n-1}$ for $n \geq 1$

2. $y_{n+1} = 2y_n + y_{n-1}$ for $n \geq 1$

3. $z_{n+1} = 2z_n + z_{n-1}$ for $n \geq 1$

4. $x_n = y_n z_n$ for $n \geq 0$

5. $y_{n+1} = y_n + z_n$ for $n \geq 0$

6. $z_{n+1} = 2y_n + z_n = y_{n+1} + y_n$ for $n \geq 0$

7. $2z_n = y_{n-1} + y_{n+1}$ for $n \geq 1$

8. $2x_n = y_{2n}$ for $n \geq 0$

9. $3x_n = z_{2n} + x_{n-1}$ for $n \geq 1$

10. $x_{n+1} - x_n = y_{2n+1}$ for $n \geq 0$

11. $x_{n+1} + x_n = z_{2n+1}$ for $n \geq 0$

12. $x_{n+1}^2 - x_n^2 = x_{2n+1}$

$$x_n(x_{n+1} - x_{n-1}) = x_{2n}$$ for $n \geq 0$

13. $y_{n+1} = z_0 + z_1 + \cdots + z_n$ for $n \geq 0$

14. $y_n^2 + y_{n+1}^2 = y_{2n+1}$

$$z_n^2 + z_{n+1}^2 = 2y_{2n+1}$$ for $n \geq 0$

15. $y_n z_{n+1} - y_{n+1} z_n = 1$ for n odd

$y_n z_{n+1} + y_{n+1} z_n = -1$ for n even

16. $y_n^2 = y_{n-1} y_{n+1} + 1$ for n odd

$y_n^2 = y_{n-1} y_{n+1} - 1$ for n even

17. $z_n^2 = z_{n-1}z_{n+1} - 2$ for n odd

   $z_n^2 = z_{n-1}z_{n+1} + 2$ for n even

18. $z_n^2 - 2y_n^2 = -1$ for n odd

   $z_n^2 - 2y_n^2 = 1$ for n even

19. $x_n^2 = x_{n-1}x_{n+1} + 1$ for $n \geq 1$

20. $\left(\dfrac{z_n - 1}{2}\right)^2 + \left(\dfrac{z_n + 1}{2}\right)^2 = y_n^2$ for n odd

   $\left(\dfrac{z_n - 1}{2}\right)^2 + \left(\dfrac{z_n + 1}{2}\right)^2 = y_n^2 + 1$ for n even

21. $\left(\dfrac{z_n - 1}{2}\right)\left(\dfrac{z_n + 1}{2}\right) = y_{2n-2} + y_{2n-6} + \cdots + \begin{cases} y_2 & \text{for } n \text{ even, } n \geq 2 \\ y_0 & \text{for } n \text{ odd}, n \geq 3 \end{cases}$

Items 20 and 21 look a little complicated, so we will unpack them. The idea is that, since each term in the z-sequence is odd, we can express it as the sum of two integers that differ by 1:

$$z = \frac{z - 1}{2} + \frac{z + 1}{2}$$

Take as an example the value $n = 5$, which is odd. Then $z = 20 + 21$. Item 20 tells us that $20^2 + 21^2 = 29^2$, while item 21 states that

$$20 \times 21 = 420 = 408 + 12 + 0 = y_8 + y_4 + y_0.$$

For the value $n = 6$, which is even, item 20 states that

$$49^2 + 50^2 = 70^2 + 1,$$

while item 21 yields

$$49 \times 50 = 2450 = 2378 + 70 + 2 = y_{10} + y_6 + y_2.$$

A Word Of Caution

If you extend the three given sequences, using items 1, 2, 3 to calculate the terms in succession, you will find that all of these relations persist. But merely checking some numerical cases will not guarantee the truth for all values n. However, it is possible to develop techniques to prove beyond all doubt that the equations listed above are true for all the infinitely many values of n.

FOUR NUMBERS

Determine sets of four distinct integers, which need not all be positive, that satisfy each one of the following properties, individually:

1. The sum of any three of the integers is a perfect square.

2. In addition to the sum of any three being square, the sum of all four of the integers is a perfect square.

3. The sum of any pair of the integers is a perfect square.

The set (0, 0, 0, 1) of four integers has the property that the sum of any number of them — 1, 2, 3, or 4 — is a perfect square. This example is pretty trivial, and besides, one of the numbers is repeated. For any collection of four integers, there are fifteen possible sums that can be obtained by adding one, two, three or all four of them. Assuming that all the four integers are distinct, we may ask what is the largest possible number of these fifteen sums that can be perfect squares.

Solution

1. Call the four numbers x, y, z, w. We have to solve in integers the system

$$
\begin{aligned}
x + y + z \quad &= a^2 \\
x + y \quad + w &= b^2 \\
x \quad + z + w &= c^2 \\
y + z + w &= d^2
\end{aligned}
$$

for some whole numbers a, b, c, d. Since

$$3(x + y + z + w) = a^2 + b^2 + c^2 + d^2,$$

it is necessary for the sum of the four squares to be divisible by 3. When we solve the system of equations, we find that

$$x = (x + y + z + w) - (y + z + w) = \frac{a^2 + b^2 + c^2 - 2d^2}{3}$$

$$y = \frac{a^2 + b^2 - 2c^2 + d^2}{3}$$

$$z = \frac{a^2 - 2b^2 + c^2 + d^2}{3}$$

$$w = \frac{-2a^2 + b^2 + c^2 + d^2}{3}$$

These numbers will be integers as long as the four squares for the sums of three add up to a multiple of 3.

Here are some possibilities for (x, y, z, w) :

$$(1, 22, 41, 58)$$

$$(1, 34, 65, 190)$$

$$(9, 34, 57, 78)$$

$$(33, 66, 97, 126)$$

2. If we also want all four of the numbers x, y, z, w to add up to a perfect square, then we also need to arrange that $a^2 + b^2 + c^2 + d^2$ is itself a perfect square. D.C. Baillie, of Toronto, gives two possibilities:

$$(120, 177, 232, 432)$$

$$(136, 264, 384, 441)$$

3. This is a much more challenging problem. Sallianne Dech, while she was a student at Sarnia Northern Collegiate and Vocational School in 1981, wrote an essay on this question. Letting x, y, z, w be the four desired numbers, we want to have:

$$
\begin{aligned}
x + y &= a^2 \\
y + z &= b^2 \\
x \quad + z &= c^2 \\
x \qquad + w &= d^2 \\
y \quad + w &= e^2 \\
z + w &= f^2.
\end{aligned}
$$

The squares must be selected so that:

$$a^2 + f^2 = b^2 + d^2 = c^2 + e^2.$$

By taking $a = 0$, Dech realized that (b, d, f), (c, e, f) were pythagorean triples and was able to find some solutions, such as

$$(88, -88, 137, 488),$$
$$(448, -448, 1073, 3152), \dots$$

Starting with the pythagorean triples, (3, 4, 5), (8, 15, 17) and (13, 84, 85), we observe that

$$51^2 + 68^2 = 17^2(3^2 + 4^2) = 17^2 \times 5^2 = 85^2$$

$$40^2 + 75^2 = 5^2(8^2 + 15^2) = 5^2 \times 17^2 = 85^2.$$

Hence

$$51^2 + 68^2 = 40^2 + 75^2 = 13^2 + 84^2 = 85^2.$$

Substituting these figures for

$$a^2 + f^2 = b^2 + d^2 = c^2 + e^2$$

leads to the following system of equations:

$$x + y = a^2 = 51^2$$
$$y + z = b^2 = 40^2$$
$$x + z = c^2 = 13^2$$
$$x + w = d^2 = 75^2$$
$$y + w = e^2 = 84^2$$
$$z + w = f^2 = 68^2$$

solving which we arrive finally at

$$(x, y, z, w) = (585, 2016, -416, 5040),$$

a set of four numbers for which, not only is the sum of any pair a square, but also the sum of all four is a square.

Chapter Three

Geometry

AREA

There are various mathematical formulae which can be used to find areas of figures in the plane, but sometimes the workings can get quite messy. Often such problems can be handled quite easily with a little insight and ingenuity combined with some elementary school mathematics. Here are some problems for you to consider.

Problems

1. In each of the following, we take a square whose side length is 1, subdivide it in some way and then ask for the area of some region inside it.

   (a) In this square, the diagonal and the line from a vertex to the midpoint of an opposite side are drawn to split the square into four regions *a, b, c, d* as shown.

   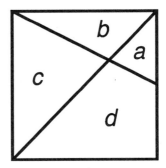

   Find their areas.

(b) The lines joining midpoints of sides and vertices are drawn as shown.

What is the area of the region marked *u*?

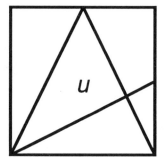

(c) Each vertex of the square is joined to the midpoint of one of the opposite sides as shown.

Explain why the region *v* is a square and determine its area.

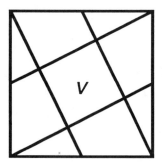

(d) Each vertex of the square is joined to the midpoint of both of its opposite sides, the segments bounding a central octagon *w*.

What is the area of *w*?

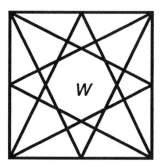

(e) Four quarter circles centred at the vertices of the square with radius 1 bound a region *s*.

Find the area of *s*.

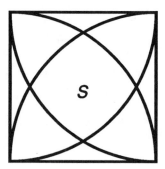

2. Two circles of radius 1 intersect in such a way that the areas marked *a* and *b* are equal. How far apart are the centres of the circle?

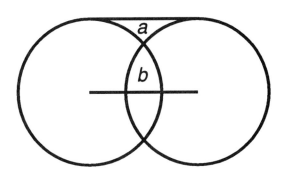

3. Each vertex of an equilateral triangle is connected to a point on the opposite edge which divides the edge in a 2 : 1 ratio, as shown in the diagram. Determine the ratio of the area of the inner triangle *t* to the area of the given triangle.

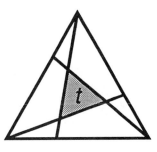

4. Here is a more difficult problem from L.J. Upton which requires some trigonometry and high school algebra. Each vertex of a regular pentagon is joined to the midpoint of one of the opposite edges, as shown in the diagram. Determine the ratio of the area of the inner pentagon *p* to the area of the given pentagon.

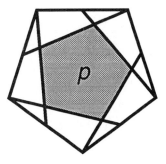

5. With a little bit of trigonometry, it can be shown that, for any circle, the ratio of the areas of the regular circumscribed and inscribed polygons is $\sec^2 \dfrac{180^\circ}{n}$, where n is the number of sides.

However, when $n = 3$ (triangle), $n = 4$ (square) and $n = 6$ (regular hexagon), there is a much simpler way to find the ratio. What is it in each of these cases?

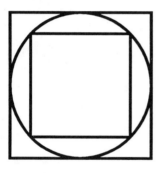

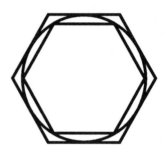

6. For the five circles pictured, find a straight line (determined from properly constructible points) which will split their combined area into two equal parts.

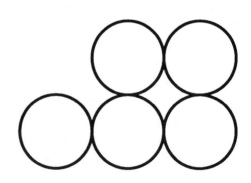

7. For the tower of nine squares pictured, determine a line passing through the point *P* which will split the area of the nine squares into two equal parts.

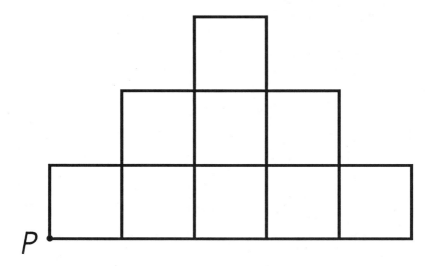

8. The inverted T-shaped region pictured is made up of five equal squares. Through the point *P* draw a line which divides the area of the region into two equal parts.

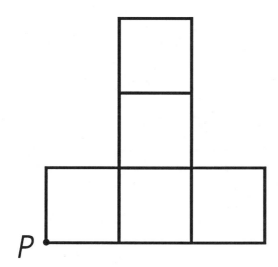

9. The letter E in the diagram is 10 cm. high, 6 cm. wide at the bottom and top, 4 cm. wide at the centre and 2 cm. thick.

Show how it can be cut with three straight scissor cuts into four pieces which can be reassembled without folding to form a solid square.

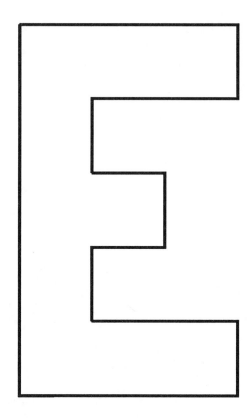

10. Three corners of a square are clipped off to form a pentagon as shown.

Cut the pentagon with two straight scissor cuts into three pieces which can be assembled into a solid square.

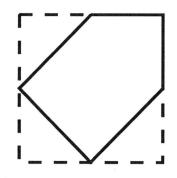

11. A regular hexagon and an equilateral triangle have the same perimeter. What is the ratio of their areas?

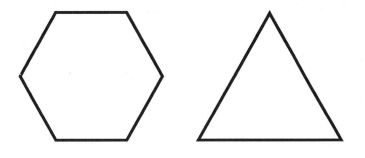

12. Determine the area of the region *t* in the diagram, given that the areas of the three triangles are as indicated.

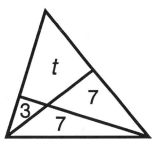

Hints

1. (a) Triangles *a* and *c* are similar. The combined area of any two adjacent subdivisions in the figure is easy to find.

(b) Look for similar triangles. Find the area of that part of the square lying outside of *u*.

(c) Consider a rotation of 90° about the centre of the square. What happens to the region marked *v*? The subdivision of the square contains four small triangles; imagine that these triangles are cut out and moved to new positions.

(d) Be careful not to assume that the octagon *w* has all of its angles equal. Subdivide the square into 9 equal smaller squares in the usual way; the central octagon consists of the central smaller square augmented by four congruent triangles, whose areas can be determined.

(e) Determine the area of that part of the square lying outside of *s*. This is equal to four times the area shaded in the diagram.

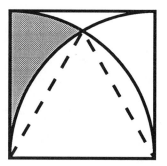

2. Form a rectangle, the sides of which are the tangent line, the line joining the centres, and the radii joining the respective centres to points of tangency. How is its area related to the area of a quarter circle?

3. Identify angles of 60° and look for similar triangles.

4. This is not an elementary problem and requires knowledge of trigonometry. In particular, one might have to know the law of cosines and the cosine of 72°.

5. In each case, rotate the inner figure so that its vertices coincide with the points of tangency between the circle and the sides of the outer figure.

The outer figure now consists of the inner figure supplemented by triangular flaps; the inner figure can be decomposed into regions congruent to these triangular flaps.

6. One way to do this is to determine a point, any line through which bisects the combined area of the four right hand circles.

7. The line through P will exit the diagram through a side of one of the component squares; determine which side this is.

8. Find the two lines through P which divide the combined area in the ratio 3 : 2.

9. Work out what the length of a side of the square must be; then see how you can make a cut of that length in the figure E.

10. Again, you know what the length of a side of the square must be. Make cuts of the appropriate length to split two of the right angles in the figure.

11. Partition each figure into congruent triangles.

12. There are a number of ways of approaching this one. The two lines from the vertices to the opposite sides are known as *cevians*. Either connect the end points of the two cevians or draw a third cevian through the intersection of the other two. Recall that the areas of triangles with the same heights are proportional to their bases.

Solutions and Comments

1. (a) Using the letters to denote the areas of their regions, we find that

$$a + b = \frac{1}{4}; \quad b + c = \frac{1}{2}; \quad a + d = \frac{1}{2}.$$

The triangles of areas a and c are similar, the latter having twice the linear dimensions of the former. Hence $c = 4a$. The equations can now be solved to yield:

$$a = \frac{1}{12}; \quad b = \frac{1}{6}; \quad c = \frac{1}{3}; \quad d = \frac{5}{12}.$$

(b) $p = q + r = r + s = \frac{1}{4}$. The triangles of areas r and s being similar, we find that $s = 4r$. Hence $r = \frac{1}{20}$ and $q = s = \frac{1}{5}$.

Thus, $p + q + r + s = \frac{7}{10}$ and so $u = \frac{3}{10}$.

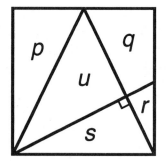

(c) A rotation of 90° counter clockwise about the centre of the square takes the line *AE* to *BF,* so *AE* must be perpendicular to *BF.* In this way, we can see that each adjacent pair of sides of the inner figure *JKLM* is perpen-dicular. This rotation also takes *K* to *L* (the intersection *CG* and *DH* goes to the inter-section of the images *DH* and *AE* under the rotation), *L* to *M, M* to *J,* and *J* to *K.* Thus *KL* = *LM* = *MJ* = *JK,* so *JKLM* is a square.

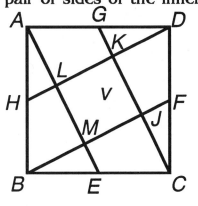

It is easy to see that the inner square has area $\frac{1}{5}$ once one rotates triangle DKG about *G* to complete the square with three vertices ALK, etc.

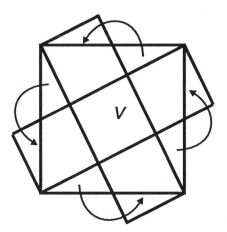

(d) We have several solutions:

Since PC is parallel to QE and C is the midpoint of AE, G is the midpoint of AN. Similarly, N is the midpoint of GR, so that G and N trisect the segment AR. Hence, B and D trisect the side AE. Therefore, the area of the central square $GINM$ is $\frac{1}{9}$.

Now consider triangle ILN. The length of its base IN is $\frac{1}{3}$, while its height is $\frac{1}{4}$ of the length of DE, namely $\frac{1}{12}$. Thus the area of triangle ILN is $\frac{1}{72}$.

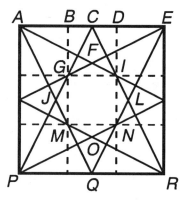

The octagon is made up of the central square and four congruent triangles, so its area is:

$$\frac{1}{9} + 4\left(\frac{1}{72}\right) = \frac{1}{6}.$$

(R. Chan)

Another solution: The octagon can be partitioned into eight congruent triangles FIK, where FK lies along a line joining the midpoints of opposite sides of the square and IK lies along a diagonal.

The length of FK is $\frac{1}{4}$.

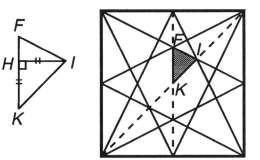

Length $FH = \dfrac{1}{2}$(Length HI)

$\qquad = \dfrac{1}{2}$(Length HK)

$\qquad = \dfrac{1}{3}$(Length FK) $= \dfrac{1}{12}$

$\qquad = \dfrac{1}{12}.$

Therefore, the length of HI is $\dfrac{1}{6}$, the area of the triangle FIK is $\dfrac{1}{2} \times \dfrac{1}{4} \times \dfrac{1}{6} = \dfrac{1}{48}$. Hence the area of the octagon is $8 \times \dfrac{1}{48} = \dfrac{1}{6}.$

(R. Cherniak)

Finally, this solution. The diagram below shows the upper right hand portion of the square. The area of triangle CKL is $\dfrac{1}{16}$. Since the areas of $\triangle CFI$, $\triangle FKI$, and $\triangle KLI$ are equal, each area is $\dfrac{1}{48}$. The area of the central octagon is equal to four times the area of the quadrilateral $FKLI$, i.e. to $4\left(\dfrac{2}{48}\right) = \dfrac{1}{6}.$

(F.V.H.)

(e) Find the area of the region in the square which lies outside of s. This is four times the area of the shaded region DBA. The region DBA is formed by putting together the 30° sector BDA and the equilateral triangle ABC, then removing

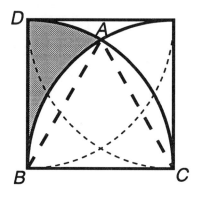

the 60° sector CBA with its centre at C. Thus, the shaded region has area $\dfrac{\sqrt{3}}{4} + \dfrac{\pi}{12} - \dfrac{\pi}{6}$. The area of region s is $\dfrac{\pi}{3} - \sqrt{3} + 1$.{A.W.W.}

2. The area of the rectangle is

$a + b + c + d$

$= 2b + c + d$

$= (b + c) + (b + d)$

$= 2\dfrac{\pi}{4}$

$= \dfrac{\pi}{2}$,

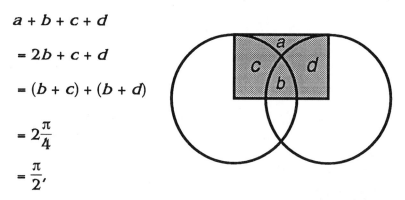

since $b + c = b + d$ is the area of a quarter circle. The area of the rectangle is equal to the distance between the centres multiplied by its height 1. Hence the centres are $\dfrac{\pi}{2}$ units apart.

3. Let x, y, z denote the lengths of segments as indicated in the diagram. Note that the inner triangle marked t is equilateral with 60° angles. From similar triangles, we obtain:

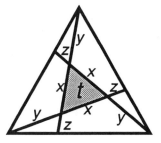

$$\frac{\frac{1}{3}}{y} = \frac{x+y+z}{1} \quad \text{or} \quad \frac{1}{3y} = x+y+z$$

$$\frac{y}{z} = 3 \quad \text{or} \quad y = 3z$$

$$\frac{2}{3} = \frac{x+z}{x+y} \quad \text{or} \quad 2y = x+3z.$$

Solving these three equations gives $x = y = 3z$ and $x^2 = \frac{1}{7}$. Since the area of similar triangles varies as the square of the linear dimensions, the area of the inner triangle marked t must be $\frac{1}{7}$.

4. Let $z = \cos 72°$. Then

$$z = 2\cos^2 36° - 1$$

$$= 2\cos^2 144° - 1$$

$$= 2(2\cos^2 72° - 1)^2 - 1$$

$$= 8z^4 - 8z^2 + 1,$$

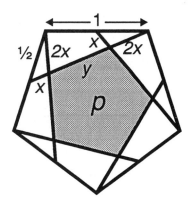

whence

$$0 = 8z^4 - 8z^2 - z + 1$$

$$= (z - 1)(2z + 1)(4z^2 + 2z - 1).$$

The first two factors on the right side being nonzero, we find that

$$z = \cos 72° = \frac{-2 + \sqrt{20}}{8} = \frac{-1 + \sqrt{5}}{4} = \frac{1}{2}(\tau - 1)$$

where τ is the "golden ratio" $\frac{1}{2}(1 + \sqrt{5})$, with $\tau^2 = \tau + 1$.

Let $u = 3x + y$, the length of the segment joining the vertex of the outer pentagon to the midpoint of the second side away from it. From similar triangles, we have $u : 1$ as $\frac{1}{2} : 2x$, so that $x = \frac{1}{4u}$.

Hence, y, the length of a side of the inner pentagon, is equal to $u - 3x = \frac{4u^2 - 3}{4u}$.

From the Law of Cosines,

$$u^2 = 1 + \frac{1}{4} + \cos 72° = \frac{1}{4}(5 + 4z).$$

The ratio of the area of the inner to the outer polygons is equal to

$$y^2 = \frac{(4u^2 - 3)^2}{16u^2} = \frac{(1 + 2z)^2}{5 + 4z} = \frac{\tau^2}{3 + 2\tau} = \frac{\tau + 1}{2\tau + 3}$$

$$= \frac{7 + \sqrt{5}}{22} = \frac{2}{7 - \sqrt{5}}$$

Another solution: *O* is the common centre of the two pentagons; *AB* and *BC* are two sides of the outer pentagon; *D* is the midpoint of *BC*; *E* is the midpoint of *BF*. Take *BD* and *DC* to be one unit in length. Let θ be the common value of angles *DOP* and *ADE*.

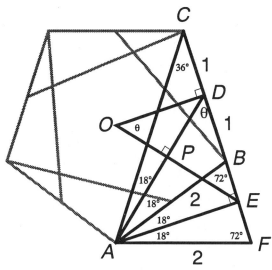

$$\tan \theta = \frac{EA}{DE} = 2\frac{EA}{CA} = 2 \sin 36°$$

The area ratio of outer to inner pentagon is the square of the ratios of the distances from *O* to the sides of the outer and inner pentagons;

$$\left(\frac{OD}{OP}\right)^2 = sec^2 \theta = 1 + 4 \sin^2 36°$$

$$= 1 + 2(1 - \cos 72°)$$

$$= 3 - \frac{BF}{BA} = 3 - \frac{FA}{CF}$$

Now, *FA* = 2. By similar triangles, $\dfrac{DF + 1}{2} = \dfrac{2}{DF - 1}$, whence *DF* = √5, and *CF* = 1 + √5. Thus, the ratio is $\frac{1}{2}(7 - \sqrt{5})$. [A.W.W.]

5. For the triangle, square, and hexagon, the ratios of the circumscribed areas to inscribed areas are, respectively,

$$4 : 1, 8 : 4 = 2 : 1, \text{ and } 24 : 18 = 4 : 3,$$

as the diagrams show.

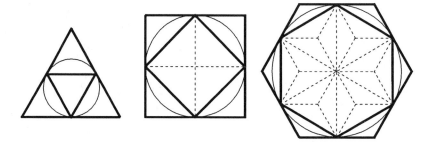

6. Here are two possibilities:

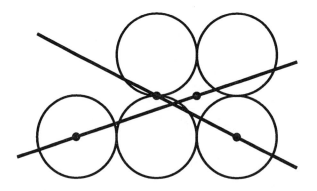

7. The line *PA* has the area of 5 squares below it, while the line *PB* has the area of 3 squares below it. Thus, the area of *PAB* is equal to the area of 2 squares. We have to choose *Q* so that *AQ* : *QB* as 1 : 4 to get a line with the area of $4\frac{1}{2}$ squares below it.

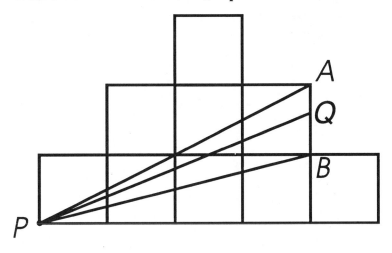

8. Each of the dotted lines divides the area in the ratio 2:3. Thus, the solid line, where *Q* is the midpoint of *RS*, is the required line.

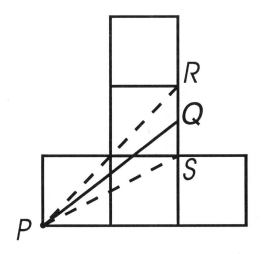

9. The total area of the letter is 40 cm^2, so each side of the square has length $\sqrt{40}$. This is the diagonal length of a 2×6 rectangle, so a natural first cut is that separating A and B in the diagram. Following this by a cut severing C and D gives us 4 edges of the desired square. Finally, sever B and C, and rearrange.

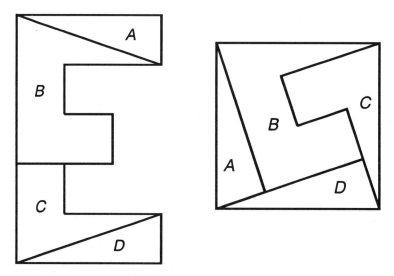

10. For convenience, let each side of the original square have length 4, so that the pentagon has three sides of length $2\sqrt{2}$ and two sides of length 2; its area is 10. We seek a length of $\sqrt{10}$ to serve as a side of the desired square. Two possibilities present themselves: a line from A to the midpoint of one of the three equal sides of the pentagon, and a line from one of the

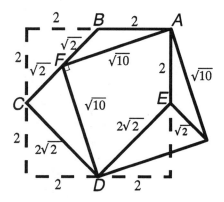

vertices B, C, D, E to the midpoint of one of the three equal sides. The solution is given in the diagram.

11. By creating an equilateral triangle of four smaller equal triangles and a regular hexagon of six of the smaller triangles (as pictured), we see easily that the ratio of areas is 3:2.

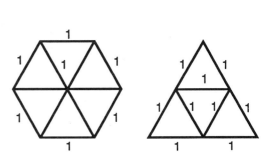

12. The correct answer is 18.

There are at least two ways of solving problems of this type.

In the figure at the right, we want to obtain $r + s$, given v, c, and p. We have

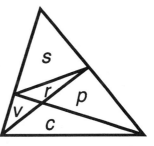

(1) $$\frac{s + r + v}{p + c} = \frac{s}{p + r}$$

(2) $$\frac{v}{c} = \frac{r}{p}.$$

Using (2) to eliminate r from (1) yields an equation for s:

$$[c(c^2 - pv)]s = pv(p + c)(c + v).$$

In the present situation, $p = c = 7$, $r = v = 3$ and $s = 15$.

Alternatively, in the figure below we have

$$\frac{u}{v} = \frac{p+q}{a+b} = \frac{p+q}{c};$$

$$\frac{p}{q} = \frac{a+b}{u+v} = \frac{c}{u+v}.$$

Massaging these a little to eliminate u, then q, yields

$$q(c^2 - pv) = pv(p + c)$$

$$u(c^2 - pv) = pv(v + c)$$

whence $u + q$ can be found.

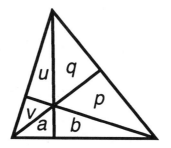

REGULAR POLYGONS FROM SQUARES

You are given a square piece of paper. Without having recourse to any measuring instruments (in particular, neither ruler nor compasses), show how, by folding and tearing off the excess paper, you can obtain from this piece of paper:

1. A regular octagon, an eight-sided figure with all sides equal and all angles equal to 135° (like a STOP sign).

2. A regular hexagon, a six-sided figure with all sides equal and all angles equal to 120°.

Hints

1. For the octagon, all you have to do is turn down the four corners of the square by the right amount and rip them off. The trick is to figure out how big a dog-ear you need. The diagonals of the square pass through the midpoints of two opposite sides of the octagon. Assuming that the square has side length 1, we see that the distance between two opposite sides of the octagon should also be 1. The diagonal of the square has length $\sqrt{2}$, so we can figure out by how much the diagonal has to be folded back.

2. The hexagon problem is a little harder. Begin with the observation that a long diagonal of the hexagon (joining opposite vertices) is twice the length of one of the sides. One strategy is to make a sketch in which a long diagonal of the hexagon is the line joining the midpoints of opposite sides of the square. Sketch in the rest of the hexagon, labelling the size of the various angles that occur.

Solutions

1. The Octagon.

In the diagram, the square, *ABCD*, has sides of length 1, and the octagon we want is *PQRSTUVW*.

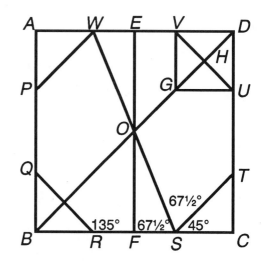

Since an isosceles right triangle has sides in the ratio $1 : 1 : \sqrt{2}$, we can obtain the following relations:

$$BF = \frac{1}{2}$$

$$DO = OB = \frac{\sqrt{2}}{2} = \frac{1}{\sqrt{2}}$$

$$HD = \frac{\sqrt{2} - 1}{2}$$

$$DG = UV = \sqrt{2} - 1$$

$$BG = BC = 1$$

$$BS = BO = \frac{1}{\sqrt{2}}$$

OS bisects ∠*FOC*.

Here are some possible processes:

(a) Fold to obtain the diagonal *BD*. Fold side *BC* up to the diagonal *BD* along the angle bisector of ∠*DBC*, so that *C* falls on *G*. Then fold *D* over to *G* to determine the side *UV*. The remaining sides of the octagon can be obtained in the same way. (J.P.F.)

(b) By a couple of diagonal folds, determine the midpoint of the square. Fold *BC* to the diagonal *BD* and determine *S* as the point that falls on *O*. The remaining vertices of the octagon can be found in the same way. {D.C.B. & L.G.H.}

(c) Fold to obtain the lines *EF* and *AC*; *O* is the intersection of these lines. Fold *OF* over to *OC* to bisect ∠*FOC*. This determines *S* and the remaining vertices of the octagon can be found in the same way. (R.C.M.)

(d) Fold *BA* over to *CD* along *EF*; then fold *FC* up to *ED*. Now fold along *OD* so that everything is on top of triangle *OED*. Now fold *OD* on to *OE* (along *OV*), and then fold D back on to *EO*, producing the fold *VH*. Now unfold the whole square, and you will find folds along *VU*, *TS*, *RQ* and *PW*. (F.V.H.)

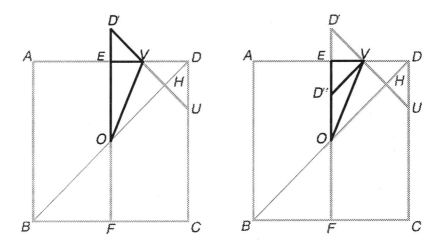

2. The Hexagon.

(a) Folding twice, obtain three horizontal folds that split the square into four equal rectangles. As indicated in the diagram, the left end of two of the folds are vertices A and B of the hexagon. By means of a fold through A, bring B down to the point F on the bottom edge of the square. The fold through A meets the top horizontal fold at D. Folding along the central horizontal fold yields the points C and E. $ABCDEF$ is the required hexagon. [R.C.M.]

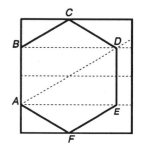

(b) Let the square be $PQRS$. Divide the square into four congruent rectangles by folding it twice horizontally. By means of a fold through P, take Q to a point T on the midline of the square. Note that $AD = PT = PQ = CF$. Make a vertical fold through T and determine the hexagon $ABCDEF$ as indicated in the diagram. [J.G.F.]

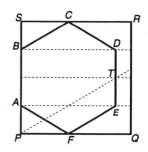

(c) Fold a square into sixteenths as indicated. B is found by a fold through A to make OA equal to AB. Continue on. [J.P.F.]

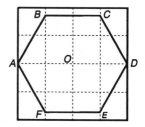

SOME SURPRISING ANSWERS

In each of the following geometric problems, your first impression of the correct solution may not be right. Think about each of them carefully !

1. The bottom of a ladder leaning against a wall is slid along the floor away from the wall. What is the path traced out by a point halfway up the ladder?

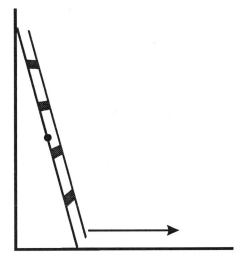

2. As the larger circle turns, a smaller circle within it rolls against the inner side without slipping. The larger circle has radius twice that of the smaller circle. What is the path traced by a fixed point on the circumference on the inner circle?

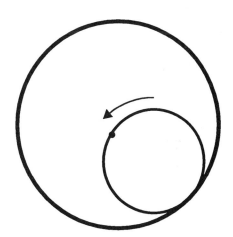

3. *A* and *B* are two distinct points on a page. Describe the set of points, *P*, whose distance from *A* is twice the distance from *B*.

•   •
A   *B*

Comments

1. Imagine the foot of the ladder right against the wall. As it begins to move out, what do you think the initial direction of the midpoint of the ladder is?

2. You might perform an experiment with two circles or discs.

3. Observe that the farther away a point *P* is from both *A* and *B*, the closer the ratio of the distances from *P* to *A* and *B* gets to 1. This rules out the possibility that the required set is a straight line extending indefinitely far in both directions.

Solutions and Comments

1. The path (or locus, to use the technical term) is a quarter circle the centre of which is where the wall and floor meet and whose radius is half the length of the ladder. The key to seeing this is the fact that the midpoint of the hypotenuse of a right triangle is the same distance from all three vertices of the triangle.

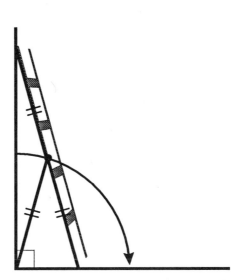

2. The locus is the straight line joining the two positions where the fixed point is at the point of tangency between the larger and the smaller circles. In the National Museum of Science and Technology in Ottawa, there is a model of the hypocycloidal engine design by Matthew Murray of Leeds, England, in 1802 that exploits this fact to convert the up and down motion of a piston into a circular motion.

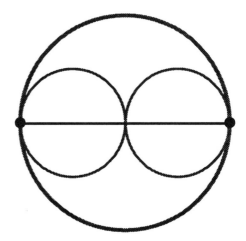

3. The locus of *P* is a circle whose diameter is a segment
   joining the point *C* on *AB*, two-thirds the way from
   A to *B*, to the point *D* on *AB* produced for which
   AB = *BD*.

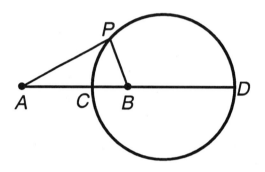

THE TWO EQUAL ANGLES

The diagram illustrates a quadrilateral *ABCD* with one of its diagonals *AC*. It is given that *AB = CD* and that the sum of the angles *DAC* and *ACB* is equal to 180° (a straight angle).

Prove that the opposite angles *B* and *D* are equal.

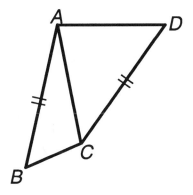

Solution

Flip the triangle ABC over so that the ends A and C of the diagonal are interchanged. This triangle in its new position with the triangle ACD will create an isosceles triangle, as BC will now form a straight line with AD. B and D are the equal angles of this isosceles triangle.

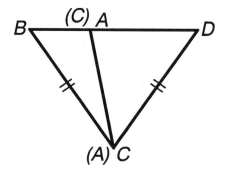

DIVIDING INTO FOUR EQUAL PARTS

The shaded figure illustrated below is bounded by a semi-circle of diameter 2 and two semicircles of diameter 1. Show how to divide it into four congruent nonoverlapping pieces.

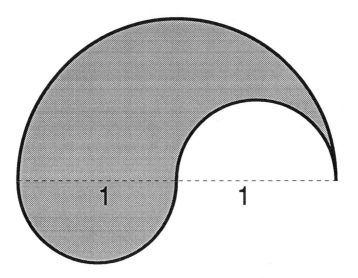

Solution

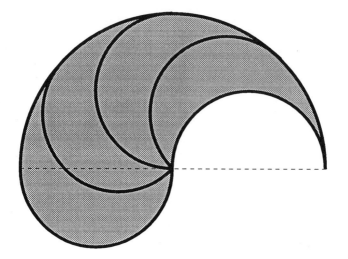

DISTANCES DETERMINED BY FOUR POINTS IN THE PLANE

Four distinct points in the plane determine 6 line segments. The line segments cannot all have the same length, so that at least two different distances are determined between pairs of points. For example, the vertices of a unit square determine 6 segments, 4 of which are sides of the square with length 1, and 2 of which are diagonals with length 2.

Find all essentially different configurations of 4 points which determine exactly two distances between pairs of them.

Hints

You have to be careful that you pick up all of the possibilities. Consider different cases: since one distance must occur at least three times, look for possibilities that it might occur five, four, or three times. In some cases, three of the equal segments might form an equilateral triangle; in others, they might emanate from the same vertex.

Solution

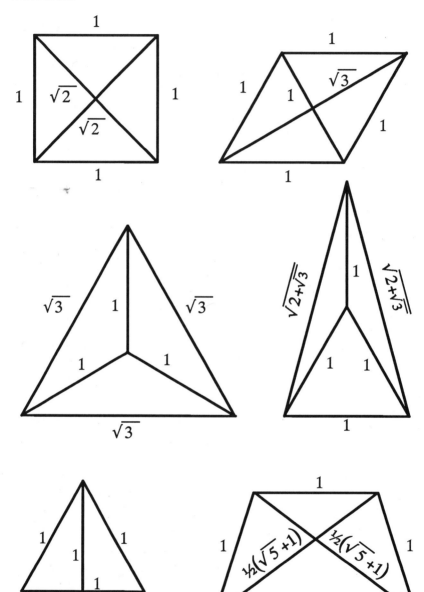

A PAPER-FOLDING PROBLEM

Probably the most successful problem I have ever posed in terms of the variation of responses received is the following paper-folding question. Take any rectangular sheet of paper—a standard $8\frac{1}{2}'' \times 11''$ sheet of typing paper will do. Suppose it is *ABCD* with *AB* longer than *BC*.

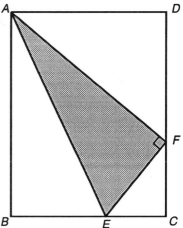

By means of a straight fold *AE* through *A*, let the vertex *B* fall on a point *F* on the side *CD*.

Perform a simple experiment to determine whether the area of the triangle *AEF* is more or less than half that of the trapezoid *AECD*.

Comments

There is no need to go for the heavy machinery on this one, defining variables and setting up equations or getting into trigonometry. This one is well within the reach of an elementary school child; all one really needs to know is the standard base and height formulae for the areas of a rectangle, triangle, and trapezoid. Note that the triangles *ABE* and *AFE* are congruent, so that *BE* = *EF* and *AB* = *AF*. Now, by making other folds in suitable places, one can find the answer to the question.

Be sure to give this one a good try before proceeding to the cornucopia of solutions which follow.

Solutions

1. Fold the side *BC* up to the side *AD* to determine the midpoint *M* of the side *CD*. If *F* lies between *C* and *M*, then triangle *AEF* covers less than half of the trapezoid; if *F* lies between *D* and *M*, the triangle *AEF* covers more than half of the trapezoid. If *M* and *F* coincide, then triangle *AEF* covers precisely half of the trapezoid.

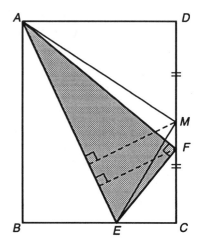

To see why this works, note that $DM = CM = \frac{1}{2}CD$ and that:

$$\text{Area } (MAD) + \text{Area } (MEC) = \frac{1}{2}DM \times AD + \frac{1}{2}CM \times EC$$

$$= \frac{\frac{1}{2}CD \times AD + \frac{1}{2}CD \times EC}{2}$$

$$= \frac{CD \times \frac{1}{2}(AD+CE)}{2}$$

$$= \frac{\text{Area } (AECD)}{2}.$$

Hence

$$\text{Area } (MAE) = \text{Area } (AECD) - \text{Area } (MAD) - \text{Area } (MEC)$$

$$= \frac{\text{Area } (AECD)}{2}.$$

Now compare triangles *AEF* and *AEM*. They have a base *AE* in common, so their areas bear the same ratio as their heights from their respective vertices *F* and *M* to the base *AE*. This height is larger the farther *F* is from *C* and closer to *D*.

2. Fold edge *AB* over a line parallel to itself so that *B* falls on *E* and *A* on *G*. Let the fold be *HK*.

Now fold *HK* over *EG*. If *HK* goes beyond *CD*, then the area of the triangle exceeds half of the area of the trapezoid; if *HK* falls short, the area of the triangle is less than half the area of the trapezoid.

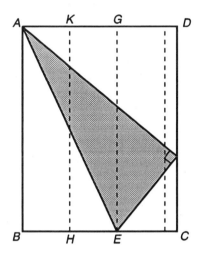

The key to this one is to note that the area of the rectangle *ABHK* is equal to the area of the triangle *ABE*, so that the question is equivalent to determining whether the flap *HEGK* covers more or less than half of the rectangle *HCDK*.

We note for later reference that the triangle covers less or more than half of the area of the trapezoid according as $BE < 2EC$ or $BE > 2EC$.

3. Fold the corner triangle *CEF* along the line *EF* and see where *C* falls. If *C* goes inside the triangle *AEF*, then the triangle covers less than half the trapezoid; if *C* goes outside triangle *AEF*, then the triangle covers more than half. Finally, if *C* falls on *AE*, then the triangle covers exactly half of the trapezoid.

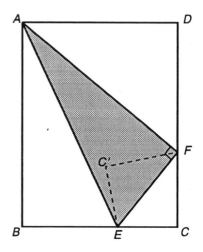

To justify this, note that *BE = EF < 2EC* if and only if sec ∠*FEC* < 2 if and only if ∠*FEC* < 60°. But this occurs exactly when

$$\angle AEF = \frac{1}{2}\angle BEF > \frac{120°}{2} = 60°$$

and *C* falls inside triangle *AEF*.

4. Fold along *RC* so that *D* falls on *S* on the edge *AB*. Also make a fold along *EG* parallel to *AB*. Then, Area(*AECD*)

= Area (*AEG*)

   + Area (*CDR*)

   + Area (*TCE*)

   – Area (*RTG*)

= 2Area (*AEF*)

   + Area (*TCE*)

   – Area (*RTG*).

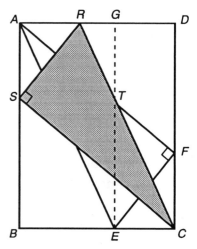

Thus, $\frac{1}{2}$Area $(AECD)$ > Area (AEF) if and only if Area (TCE) > Area (RTG), which occurs when T is closer to AD than BC (a fact that can be checked by making another fold).

5. Make a fold along DU so that C falls on V, a point on AB. $UE < EC$ if and only if triangle AEB covers less than half of the trapezoid $AECD$ when folded over.

The reader might try to find other experiments.

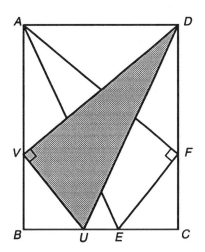

Coda

The configuration for the triangle to cover exactly half the area of the trapezoid is interesting. In this case, $EC = \dfrac{BE}{2}$, F is the midpoint of CD, and we have 60° angles at E. Hence $AB : BC$ as $2 : \sqrt{3}$.

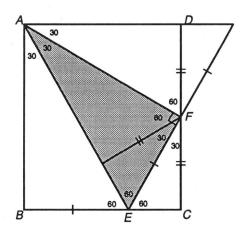

REFLECTING SPHERE

The Greeks knew that every ray of light parallel to the axis of a parabolic mirror is reflected through a single point on the axis—the focus. It is said that Archimedes took advantage of this to use sunlight in setting fire to enemy ships in Syracuse harbour; modern energy-conservers can do the same to solar-cook hamburgers. The topic of burning mirrors was quite hot among early Islamic mathematicians; some wondered about using a hemisphere instead of a paraboloid. Len Berggren of Simon Fraser University recently studied a thirteenth-century manuscript by Datrumus, who wanted to know through what points on the axis of a hemisphere the reflected rays of a beam of light parallel to the axis would pass.

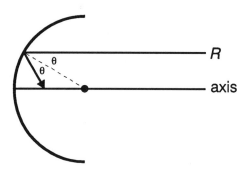

The figure shows the path of a typical ray R. Datrumus knew that the incoming and reflected rays made the same angle with a radius of the hemisphere (since any radius meets the circumference at right angles).

Problems

1. What is the path of a ray parallel to the axis that strikes the hemisphere at a point, the radius through which makes an angle of 45° with the axis?

2. Describe the path of at least one ray whose reflected ray passes through the point at which the axis meets the hemisphere.

3. Through which points on the axis will reflected rays from a beam of light parallel to the axis pass?

Solutions

1. If R is the radius of the hemisphere, the ray will come in along a line distant $\dfrac{R}{\sqrt{2}}$ from the axis, strike the hemisphere and reflect in a direction perpendicular to the axis, striking the axis at a point distant $\dfrac{R}{\sqrt{2}}$ from the centre. (If the hemisphere was actually a sphere, the path of the ray would be an inscribed square.)

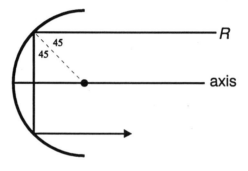

2. **If the ray passes through the end of the axis on the first reflection, the configuration is illustrated below:**

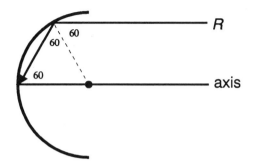

If two or three reflections are required, then we have these configurations:

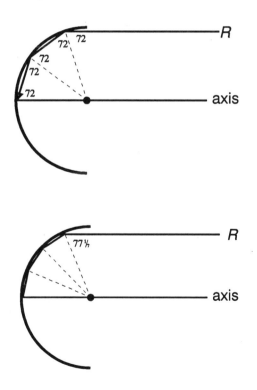

3. In the diagram below, let x be the distance between the ray's path and the axis, y the distance between the centre and point where the ray crosses the axis, and r the radius of the circle. Since $x = r \sin \theta$ and $y = \dfrac{r}{2} \sec \theta$,

$$y^2 = \frac{r^2}{4 (1 - \sin^2 \theta)} = \frac{r^4}{4 (r^2 - x^2)}$$

i.e., $$y = \frac{r}{2}\left(1 - \frac{x^2}{r^2}\right)^{-\frac{1}{2}}$$

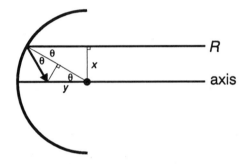

As x increases from 0 to $r\dfrac{\sqrt{3}}{2}$, y increases from $\dfrac{r}{2}$ to r. If x is greater, then the ray reflects more than once before crossing the axis, but never closer to the centre than $\dfrac{r}{2}$.

When the incoming ray is very close to the axis, the focal distance from the centre can be taken as $\dfrac{r}{2}$, as is common in applications.

TRIANGLES AND MEDIANS

Let ABC be a triangle. The three lines AD, BE, CF that join the vertices to the midpoints of the opposite sides are called its *medians*. The medians intersect in a single point known as the *centroid*.

The segment FE is parallel to and equal to half the length of the side BC. The segments DE and FD are similarly related to the respective sides BA and AC.

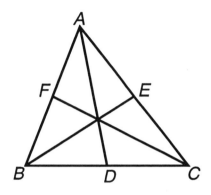

Problems

1. Explain why the length of the median AD is less than the average of the lengths of the sides AB and AC.

2. The triangle DEF is similar to and has half the linear dimensions of the triangle ABC. Accordingly, the area of $\triangle DEF$ is one-quarter of the area of $\triangle ABC$. Now construct a new triangle whose side lengths are equal to the lengths of the three medians AD, BE, CF. Explain why the sum of the areas of this new median triangle and the triangle DEF is equal to the area of triangle ABC.

3. Determine triangles whose side lengths and median lengths are all positive integers.

Comments

1. Another way to formulate the question is to ask for an argument that double the length of the median AD is less than the sum of the lengths of sides AB and AC. The triangle inequality states that the length of any side of a triangle is less than the sum of the lengths of the other two sides. Accordingly, one might ask whether there is a triangle whose side lengths are the lengths of AB, of AC, and of twice the median AD.

2. This problem does not have to be done using a formula. A construction proof, in which additional lines are drawn and the figure partitioned, is possible.

3. To do this problem, you will, of course, need the formula for the lengths of the medians of a triangle. The derivation will be left to those readers with a high school background in mathematics. Let a be the length of the side opposite vertex A, u be the length of the median from A. Let b and v be the corresponding lengths for vertex B, and c and w for vertex C. Then

$$4u^2 = 2b^2 + 2c^2 - a^2$$

$$4v^2 = 2c^2 + 2a^2 - b^2$$

$$4w^2 = 2a^2 + 2b^2 - c^2.$$

Observe that, if we start with the identity $2(p + q)^2 + 2(p - q)^2 = 4p^2 + 4q^2$ and let $(a, b, c) = (p + q, p - q, 2q)$, then the median to side c has length p. Note also that if $(a, b, c; w, v, u)$ is a side-median vector, then so also is $(4w, 4v, 4u; 3a, 3b, 3c)$. (This fact will also help out in problem 2 if you want an algebraic argument.) (D.C.B. and L.G.H.)

To avoid trivial considerations, we can require that the greatest common divisor of the integers *a, b, c, u, v, w* is 1.

Solutions

1. Make a rotation of triangle *ACD* through
   180° about the point
   D so that *C* falls on *B*
   and *A* goes to a point
   X. *A, D, X* are collin-
   ear; the length of *AX*
   is twice that of *AD*,
   and *ABX* is a triangle
   having the side *AX* of
   length necessarily
   less than the sum of
   the lengths of the
   other two sides,
   $AB + BX = AB + AC$.

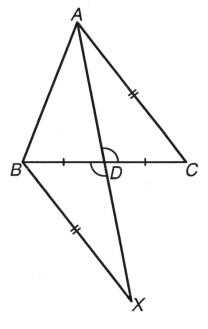

2. Let us dispose of the more mundane technical proofs first and leave the more elegant approaches (like dessert) to the end. If *A* is the area of a triangle with side lengths *a, b, c*, then a high school result called Heron's formula leads to

$$A^2 = \frac{2(a^2b^2 + a^2c^2 + b^2c^2) - (a^4 + b^4 + c^4)}{16}.$$

Using the formulae for the median lengths *u, v, w*, we find that

$$u^2v^2 = \frac{1}{16}(2b^2 + 2c^2 - a^2)(2c^2 + 2a^2 - b^2)$$

$$= \frac{1}{16}(5a^2b^2 + 2a^2c^2 + 2b^2c^2 - 2a^4 - 2b^4 + 4c^4)$$

with analogous expressions for u^2w^2 and v^2w^2, so that

$$u^2v^2 + u^2w^2 + v^2w^2 = \frac{9}{16}(a^2b^2 + a^2c^2 + b^2c^2).$$

Also

$$u^4 = \frac{1}{16}(a^4 + 4b^4 + 4c^4 + 8b^2c^2 - 4a^2b^2 - 4a^2c^2)$$

with analogous expressions for v^4 and w^4, so that

$$u^4 + v^4 + w^4 = \frac{9}{16}(a^4 + b^4 + c^4).$$

Applying Heron's area formula with u, v, w in place of a, b, c, yields the result that the area of the median triangle is $\frac{3}{4}$ that of the original triangle.

If we knew in advance that the area of the median triangle bears a fixed ratio to that of the original triangle, we would have this short argument. Note that if the original triangle is represented by the side-median vector $(a, b, c; w, v, u)$, then the median triangle is represented by $(w, v, u; \frac{3}{4}a, \frac{3}{4}b, \frac{3}{4}c)$. Repeating the process of forming triangles from medians leads to the triangle with vector $(\frac{3}{4}a, \frac{3}{4}b, \frac{3}{4}c; \frac{3}{4}w, \frac{3}{4}v, \frac{3}{4}u)$ [L.G.H.].

The third triangle is similar to the first with linear dimensions $\frac{3}{4}$ those of the first; so its area is $\frac{9}{16}$ that of the first. The ratio of the area of the second to that of the first is the geometrical mean of this, namely $\frac{3}{4}$.

A diagrammatic proof (D.C.B.):

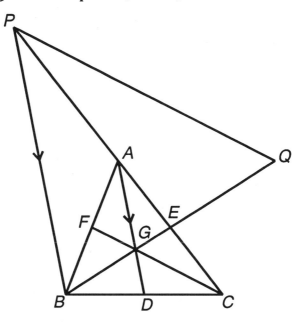

In the diagram, $AP = AC$, BP is parallel to DA and $BE = EQ$. Since the medians trisect each other, $GE = \frac{1}{3}BE = \frac{1}{3}EQ$. Also $CE = \frac{1}{3}PE$ and $\angle GEC = \angle QEP$, so triangles GEC and QEP are similar. It follows that PQ is parallel to CF. The lengths of the sides of the triangle PBQ are double the lengths of the medians, and so it has four times the area of the median triangle. Since PE is a median of triangle PBQ and $EA = \frac{1}{3}EP$, the centroid of triangle PBQ is A. We now note that the area of triangle ABE is half that of triangle ABC and one-sixth that of triangle PBQ, and so two-thirds that of the median triangle. Thus, the area of triangle ABC is four-thirds that of the median triangle.

In the following diagramatic solution, *DG* is equal and parallel to *BE*; *GA* is equal and parallel to *CF.* Taking account of numerous pairs of parallel lines, one can show that the small triangles numbered from 1 to 12 are equal in area. Six of them partition triangle *ADG* and eight of them partition triangle *ABC.* (R.C.M.)

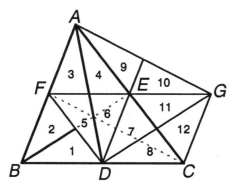

In a similar approach, in the diagram below, the median triangle is congruent to *BEH.* We find that
$\Delta FEH = \Delta AFE$; $\Delta BFH = \Delta BDF$; $\Delta BEF = \Delta DEF = \Delta DCE$.

Hence,
$\Delta BEH + \Delta DEF = \Delta FEH + \Delta BFH + \Delta BFE + \Delta DEF$
$= \Delta AFE + \Delta BDF + \Delta DEF + \Delta DCE = \Delta ABC$.

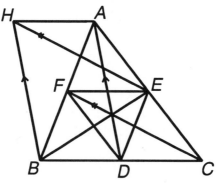

(L.G.H.)

Thus, by these various arguments we deduce that the median triangle has three-quarters of the area of triangle ABC and that the sum of the areas of the median triangle and triangle DEF is equal to the area of triangle $\triangle ABC$.

3. Here are some possibilities:

(136, 170, 174; 127, 131, 158)

(254, 262, 316; 204, 255, 261)

(284, 926, 1058; 435, 621, 984)

(290, 414, 656; 142, 463, 529)

(466, 510, 884; 208, 659, 683)

(932, 982, 1614; 515, 1223, 1252)

(1138, 1280, 1762; 831, 1338, 1431).

MAKING TRACKS

The great Canadian landscape artist, Rachel S. Chapter, was in the process of painting a winter prairie scene when her work was suddenly interrupted by a sudden and severe Saskatchewan blizzard. Incapacitated by frostbite and confined to bed with only her treasured copy of Coxeter's *Real Projective Plane* to console her, she had to let her faithful pupil, Tobor N. Verse, finish the work of art.

Titled "C. P. Snow," the painting was to portray railway tracks slicing across the limitless plain. As you can see from the reproduction below, she was able to complete only the horizon, the rails and two of the ties. Keeping in mind that Chapter is especially famous for her rigorous depiction of perspective, advise Verse on

how he should place the remaining ties on the track in order to create the impression of equal spacing.

Hints

With the exception of lines that are parallel to the horizon, lines that are parallel in the actual scene will appear to meet on the horizon. Imagine the railway track stretching across the prairie, with the ties equally spaced, so that the ties and tracks form a succession of congruent (equal) rectangles. The corresponding diagonals of these rectangles will be parallel. Now transfer the situation to the canvas.

An alternative way of looking at the situation is to consider where you would place the point on the tie which is to represent the point midway between the rails. Join the endpoint of one tie to the midpoint of the next; this line can be extended to pass through the other endpoint of a third tie.

Solution and Comments

The standard theory for perspective drawing was developed by the artist Albrecht Dürer (1471-1528). It invokes the principle that the representation of parallel lines in any direction not parallel to the horizon (or "line at infinity") should meet at the horizon.

In the drawing, we can easily find the "point at infinity" for the diagonals of the rectangle formed by the rails and adjacent ties that point towards the upper left. Join the right end of the closest tie to the left end of the next tie and extend to meet the horizon. The left end of the third tie can be found by joining this point to the right end of the second tie. We can continue on in this way as indicated in the diagram to get the remaining ties.

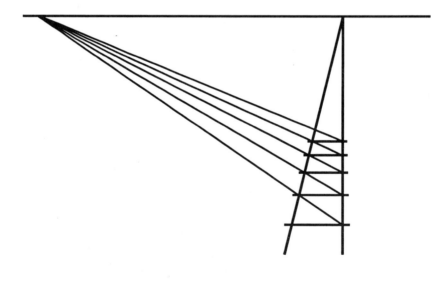

Let *AB*, *CD*, *EF* be rep-resentations of three con-secutive ties in the drawing.

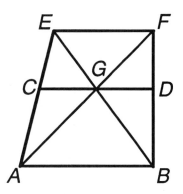

The diagonals *AF* and *BE* will intersect in the point *G* that represents the middle point of the tie rep-resented by *CD*. We shall see that *CG* = *GD* in the ac-tual diagram.

Since triangles *EGF* and *BGA* are similar, *EG* : *GB* as *FG* : *GA*. The areas of similar triangles are in the same ratio as the squares of their corresponding linear dimensions, so that

Area (ΔGEC) : Area (ΔBEA) as $EG^2 : EB^2$ and

Area (ΔGFD) : Area (ΔAFB) as $FG^2 : FA^2$.

Since ΔBEA and ΔAFB have the same base and height, their areas are equal. Also, *EG* : *GB* as *FG* : *GA* implies that *EG* : *EB* = *FG* : *FA*.

It follows that Area(ΔGEC) = Area(ΔGFD). Since ΔGEC and ΔGFD have the same heights, their bases *CG* and *GD* are equal.

This fact suggests another method of drawing in the remaining ties, once *AB* and *CD* are given. Simply join *A* and *B* to the midpoint of *CD* and extend the lines to meet the rails at *F* and *E*, which will be the endpoint of the next tie. We can apply this method to *CD* and *EF* to get the next tie, and so on.

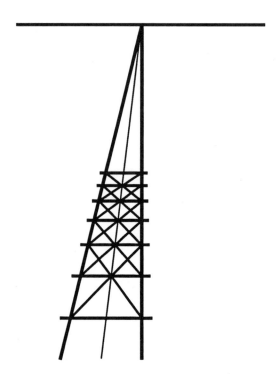

The lengths of *AB*, *CD*, and *EF* turn out to be related in an interesting way. Denote by lower case letters the lengths of various segments as follows:

AB	*a*
CD	$b = 2u$
CG, GD	*u*
EF	*c*
EG	*v*
GB	*w*

Taking note of similar triangles, we have

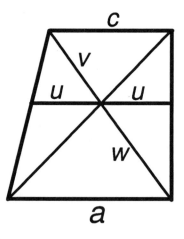

$$\frac{a}{u} = \frac{v+w}{v} = 1 + \frac{w}{v} = 1 + \frac{a}{c} = \frac{c+a}{c}$$

which leads to

$$\frac{1}{u} = \frac{1}{a} + \frac{1}{c}$$

so that

$$\frac{1}{b} = \frac{1}{2}\left(\frac{1}{a} + \frac{1}{c}\right).$$

Mathematically, this means that b is the *harmonic mean* of a and c. The lengths of the representations of consecutive ties form a harmonic sequence.

Those wishing to explore the mathematics of this further may consult the article

R.J. Duffin, "On seeing progressions of constant cross ratio," *American Mathematical Monthly* 100 (1993), 38-47.

Many of the people to whom this problem was originally posed supported one or other of the methods already discussed. W.S.A. Dale of London, Ontario, a Fine Arts graduate, wrote:

Even in its unfinished state Ms Chapter's masterpiece *C.P. Snow* clearly reveals that the artist was a pupil of the long-lamented Leonbattista Alberti. This is indicated by the convergence of the orthogonals (i.e., the railway tracks) on a single point (usually called the vanishing point) at the level of the horizon, and the strictly parallel lines of the railway ties.

Dale proposes a second method of plotting the levels of the ties, by reconstructing a side elevation of the railway tracks.

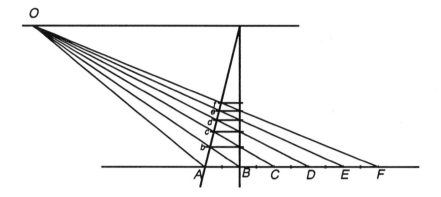

In the diagram *Abcdef...* represents the plane of the picture and *O* the eye of the observer. The ground extends from left to right, with the position of the ties at *A, B, C, D, ...* . In the picture, the successive ties are placed in positions *A, b, c, d, e, f, ...*

Two other questions that might have been asked are: (1) How far to the right of the track was the artist standing? (2) How far above the track was her point of view? Assuming that the canvas was vertical and at right angles to the tracks, and the tracks were laid to a standard gauge, a reader calculated that the artist was about 3 feet to the right of the right-hand rail and 19 feet above. "It appears that the artist must have been standing on a bridge." [G.A.C.]

Another way of proceeding is to project the equidistant points *A, B, C, D, E, ...* onto a circle with centre *O* (observer). The distance between the ties are proportional to the lengths of the chords *Ab, bc, cd, ...* [F.V.H. and R.C.M.]

Concerning this method, F. V. Harrison writes:

> May I make an observation in defence of the ratio of chords system. It is derived from

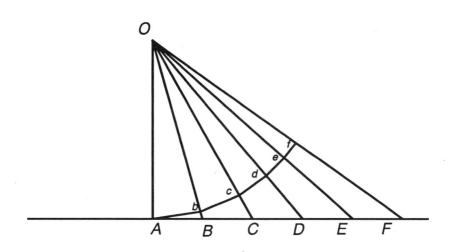

practice taught to students at the Art Gallery of Ontario around the beginning of this century.

My mother, Grace F. (née) Cooper, was among them and she used it later (in my time) when she enjoyed sketching from time to time, usually countryside scenes that took her fancy. This practice was only used in trying to capture the actual scene in the artist's view, and not in theoretical drawings used for illustrations, advertisements, etc.

She was told to grasp a pencil in one hand (the right, say) in the fist form with the pencil in the centre and one end protruding by the thumb; then to extend that arm full length towards and view the specific item or distance being captured across the protruding end of the pencil so that the one side corresponded to ...the end of the pencil and ... the tip (of the thumb) similarly corresponded to the other side. One eye is closed. If all the distances critical to the relativity of sizes portrayed are similarly measured and transferred to the canvas or sketch pad, the result is true to life as the eye sees it.

This certainly supports the circle method. Providing the observer's eye and the track are fixed, the radius of the circle is immaterial to the *ratio* of the chords. It seems to tie in with concept that the eye measures by the angle of the relative light rays.

All of which goes to show that one can use whatever model one wants, so long as the result is pleasing.

Chapter Four

Problems Needing a Systematic Search

Sometimes, the difficulty with a problem is that there may be many solutions so that it is difficult to get on a track that leads to a single one. In this case, it is often best to plunge in with a guess which is then modified until a solution is found. Sometimes, it is possible to narrow the search with a little preliminary reasoning.

In this chapter are presented some problems that illustrate this process.

TWO NUMERICAL ARRAYS

1. Fill in the numbers from 1 to 7 in place of the letters so that the two sides and the bar of the "H" all have the same sum:

J   K

L M N

P   Q

2. Fill in the numbers from 1 to 9 in place of the letters in the triangular array so that the numbers in each of the three sides of the triangle add up to 20:

R
S T
U   V
W X Y Z

Comments and Extensions

1. You might try to figure out what the common sum of the two sides and bar is. There are two letters that belong to two of these, so that if we add together the three sums, these two letters get counted twice. Thus, the total of the three sums is $1 + 2 + \ldots + 7$ plus a little bit more. This leads us to an upper bound for the sum. On the other hand, the number 7 has to go somewhere. How big must the sum of any row that contains it be?

2. If we add together the numbers in the three sides of the triangle, the numbers R, W and Z get included twice. Since you know the sum along each side and since you know the sum of all nine numbers, you can deduce the value of $R + W + Z$. Now consider where the number 5 should go. There are two possibilities: at a vertex, or in the middle of a side. Can you eliminate one of these possibilities? From this point on, it is a matter of trial and error.

Solutions

1. The sum of the three rows taken together is

$$(J + L + P) + (L + M + N) + (K + N + Q)$$

$$= 1 + 2 + 3 + 4 + 5 + 6 + 7 + L + N$$

$$= 28 + L + N.$$

$L + N$ is at most $6 + 7 = 13$ and at least $1 + 2 = 3$.

If we let S be the common sum of a row, we find that $3S$ is at most 41 and at least 31. It follows from this that S is 11, 12, or 13.

Suppose that $S = 11$. Then $L + N = 5$, and we are led essentially to two solutions:

```
  7   5          7   2
  3 6 2          1 6 4
  1   4          3   5
```

Suppose that $S = 12$. Then $L + N = 8$, and we get the solutions

```
  2   5       5   7       1   2
  7 4 1       6 4 2       5 4 3
  3   6       1   3       6   7
```

Suppose that $S = 13$. Then $L + N = 11$, and we get the solutions

```
  5   6          4   7
  7 2 4          6 2 5
  1   3          3   1
```

2. The first stage is to show that 5 must go at one of the corners of the triangle. With an argument similar to that used in Problem 1, we have that the sum of the three sides of the triangle is

$$(R + S + U + W) + (W + X + Y + Z) + (R + T + V + Z)$$

$$= 1 + 2 + 3 + 4 + 5 + 6 + 7 + 8 + 9 + R + W + Z,$$

so that $60 = 45 + R + W + Z$. Thus $R + W + Z = 15$.

Suppose, if possible, that 5 occurs in the middle of a side, say at S. Then $R + W + U = 15$. Hence, $U = Z$, which contradicts the fact that all the numbers are to be different.

Thus, 5 is at one corner. Let $R = 5$. We obtain solutions:

```
    5              5              5
  2 1            3 2            4 2
 9   8          8   7          8   6
4 3 7 6        4 1 9 6        3 1 9 7
```

```
    5              5              5
  4 1            6 3            6 2
 9   6          7   4          8   4
2 3 7 8        2 1 9 8        1 3 7 9
```

TWO PROBLEMS

1. Determine a ten-digit number such that the left digit indicates the number of zeros, the second digit the number of 1's, the third digit the number of 2's, ..., the tenth digit the number of 9's.

2. (a) Suppose that you are given a row consisting of three O's followed by three X's: O O O X X X.

   Your task is to obtain a row in which the O's and X's alternate: O X O X O X or X O X O X O. Here are the rules: you make a succession of moves, each move taking a consecutive pair of letters and moving them to either end of the row or into a pair of adjacent vacant slots. The final row need not occupy the same slots as the beginning row, but there must be no gaps between adjacent letters. Here is a solution:

   ```
   O O O X X X
     O X X X O O
   X O O X X     O
   X O     X O X O
     X O X O X O
   ```

   This is not the solution that uses the fewest moves.

   Find a shorter solution.

   (b) Is there a solution to (a) leading to O X O X O X?

   (c) Solve the same problem beginning with

   O O O O X X X X

   and

   O O O O O X X X X X

Comments

1. The task in this problem is like looking for a needle in a haystack. It is made more troublesome by its self-referencing character. One technique, often used in serious mathematics, for trying to get a solution is: start with a guess, and then use this guess to construct another guess.

   In the problem at hand, begin with any ten-digit number you want—let's call it *A*. Construct a new number *B* as follows: the first digit of *B* is the number of zeros in *A*; the second digit of *B* is the number of ones in *A*; and so on. Unless you are very lucky, the number *B* will not be the same as *A* (if *A* = *B*, then you will have found a solution). Repeat the process on *B* to get a new number *C*. If you go on doing this long enough, you will run out of numbers to get and hit a number you have seen earlier in the list. If two consecutive numbers are the same, you have a solution. If you get in a loop without two consecutive numbers the same, start over with a new value of *A*. In practice, you should hit on an answer fairly quickly.

   However, even if you get an answer, you have no assurance that there are not others to be found. To get a handle on the situation, go back to your sequence *A*, *B*, *C*, ... constructed above. What is the sum of the digits in *B*? in *C*? in ...? How do you account for this sum? There are two ways of finding the sum of the digits in the ten digit number. As an analogy, suppose that we have a pile of money to count. One way to do it is simply to add up the coins as we encounter them, now a nickel, now a dime, etc. A more systematic way is to put the coins in piles according to their denominations. Then the number of cents is 25 times the number of quarters plus 10 times the number of dimes, and so on.

Suppose the ten digits are

$$a, b, c, d, e, f, g, h, j, k.$$

There are a zeros, b ones, c twos, etc. Now write the sum of the digits in two ways, equate them and work from there.

This problem can be generalized. Let n be a positive integer. Write a sequence of $n+1$ positive integers chosen from 0 to n inclusive; suppose the sequence consists of $a_0, a_1, ..., a_n$. We require that the sequence contains a_0 zeros, a_1 ones, a_2 twos, ..., a_n n's.

A variant of the problem occurs in the book *Metamagical Themas* by Douglas Hofstadter (page 27): fill in the blanks to complete the following sentence:

> In this sentence, the number of occurrences of 0 is ____, of 1 is ____, of 2 is ____, or 3 is ____, of 4 is ____, of 5 is ____, of 6 is ____, of 7 is ____, of 8 is ____, of 9 is ____.

The sentence refers to occurrences of digits and the blanks need not be filled in by single-digit numbers. Zero (0) is never an initial digit of a number.

2. (a) One can pose the same problem with n O's and n X's where n is a positive integer. Is there a systematic way of solving the problem? What is the smallest number of moves needed?

Solution

1. Consider the 10-digit number first. Since the first digit counts the number of zeros, etc., the sum of the digits counts the number of zeros plus the number of ones, etc., i.e., the total number of digits, which is 10. Let us try to find a solution by taking a guess, and modifying that guess by making the next number's digits count the digits of each type in the first. For example, starting with 5 3 2 0 0 0 0 0 0 0 we get, in turn

$$7\ 0\ 1\ 1\ 0\ 1\ 0\ 0\ 0\ 0$$

$$6\ 3\ 0\ 0\ 0\ 0\ 0\ 1\ 0\ 0$$

$$7\ 1\ 0\ 1\ 0\ 0\ 1\ 0\ 0\ 0$$

$$6\ 3\ 0\ 0\ 0\ 0\ 0\ 1\ 0\ 0$$

which gets us into a loop. No solution here! Let us try again, starting with

$$5\ 2\ 2\ 1\ 0\ 0\ 0\ 0\ 0\ 0$$

$$6\ 1\ 2\ 0\ 0\ 1\ 0\ 0\ 0\ 0$$

$$6\ 2\ 1\ 0\ 0\ 0\ 1\ 0\ 0\ 0$$

$$6\ 2\ 1\ 0\ 0\ 0\ 1\ 0\ 0\ 0$$

The number 6210001000 satisfies the desired condition. However, there may be other solutions. To see that there are not (at least for 10-digit numbers), let us look at the general situation. Suppose that we have $n+1$ numbers $a_0, a_1, a_2, ..., a_n$ with a_0 zeros a_1 ones a_2 twos, and so on. The sum of the digits $(n+1)$ can be represented in two ways:

$$a_0 + a_1 + a_2 + \cdots + a_n$$

$$= 0a_0 + 1a_1 + 2a_2 + 3a_3 + \cdots + na_n.$$

Rearranging the terms in this equation gives

$$a_0 = a_2 + 2a_3 + 3a_4 + 4a_5 + \cdots + (n-1)a_n.$$

Now, a_0 cannot be 0 (why?). Suppose that a_0 is the positive integer r.

Suppose that $a_0 = 1$; then we must have $a_2 = 1$, $a_3 = a_4 = \cdots = 0$. Thus there are one zero, one two, and two ones (so $a_1 = 2$); this gives $n = 3$ and the sequence (1 2 1 0).

Now let $a_0 = r \geq 2$. Then a_r is at least 1.

We have that $r = a_0$ is equal to a sum that includes $(r-1)a_r$ along with other terms equal to at least 0. Hence, r is at least as big as $(r-1)a_r$. Put another way, we see that a_r is no bigger than $\dfrac{r}{r-1}$.

If $r = 2$, then $a_r = 1$ or $a_r = 2$.

If $r > 2$, then a_r must be 1.

Consider $r = 2$, $a_2 = 1$. This cannot occur, since each of the terms $2a_3$, $3a_4$, ... either exceeds 1 or is zero. Thus, if $r = 2$, then $a_2 = 2$ and $a_3 = a_4 = \cdots = 0$. a_1 could be either zero or one, yielding the possibilities (2 0 2 0) or (2 1 2 0 0).

Finally, let $r \geq 3$. Then $a_r = 1$, so that

$$r = a_2 + 2a_3 + \cdots + (r-1) + ra_{r+1} + \cdots$$

Subtracting r from each side yields

$$0 = (a_2 - 1) + \cdots$$

so that we must have $a_2 = 1$, $a_3 = a_4 = \cdots = 0$.

There must be 1 r, 2 ones (a_2 and a_r), 1 two (a_1) and r zeros.

Checking out for various values of r yields the sequences

$r = 3$	(3 2 1 1 0 0 0)	$n = 6$
$r = 4$	(4 2 1 0 1 0 0 0)	$n = 7$
$r = 5$	(5 2 1 0 0 1 0 0 0)	$n = 8$
$r = 6$	(6 2 1 0 0 0 1 0 0 0)	$n = 9$

and so on. Thus, for each value of n exceeding 5, there is exactly one solution.

Here is the solution to the Hofstadter problem:

The completed sentence with blanks filled in reads as follows:

> In this sentence, the number of occurrences of 0 is 1, of 1 is 7, of 2 is 3, of 3 is 2, of 4 is 1, of 5 is 1, of 6 is 1, of 7 is 2, of 8 is 1, of 9 is 1.

Suppose that k is the maximum number of digits of a number filling the blanks. Then the total number of digits used altogether does not exceed $10 + 10k = 10(k + 1)$. Let n be one of the k digit numbers. Then $n \geq 10^{k-1}$ and some digit must occur n times. Hence $10(k + 1) \geq 10^{k-1}$ with the result that $k + 1 \geq 10^{k-2}$ or $k \leq 2$. Hence, the blanks must be filled with one-and two-digit numbers.

Since the total number of occurrences of all the digits is not greater than 30, there are at most two two-digit numbers. In fact, one can argue that there is at most one two-digit number, so that the total number of occurrences of digits is 21 (10 in the statement of the problem, 2 in the two-digit blank and 1 in each of the remaining blanks). But one finds that this too is not possible. Thus all the blanks are filled by single-digit numbers. The sum of the digits in the blanks is 20 and the sum of all the digits involved is 65. Now narrow down to the solution.

2. (a) A shorter solution

```
O O O X X X
   O X X X O O
   O X X      O X O
      X O X O X O
```

(b)

```
O O O X X X
O      X X X O O
O X O X X      O
O X O      X X O
O X O X O X
```

(c) Four 0's and four X's:

```
O O O O X X X X
O      O X X X X O O
O X X O      X X O O
O X X O X O X      O
   X O X O X O X O
```

(4 moves)

```
O O O O X X X X
O O O      X X X O X
O O O X X      X O X
O      X X O O X O X
O X O X      O X O X
O X O X O X O X
```

(5 moves)

Five 0's and five X's

```
0 0 0 0 0 X X X X X
0       0 0 X X X X X 0 0
0 X X 0 0 X X     X 0 0
0 X X 0     X 0 X X 0 0
0 X X 0 X 0 X 0 X       0
    X 0 X 0 X 0 X 0 X 0
```

(5 moves)

```
0 0 0 0 0 X X X X X
0 0 0 0     X X X X 0 X
0 0 0 0 X X X     X 0 X
    0 0 X X X 0 0 X 0 X
X 0 0 0 X X     0 X 0 X
X 0 0     X 0 X 0 X 0 X
    0 X 0 X 0 X 0 X 0 X
```

(6 moves)

Believe it or not, there is a general way of starting with *n* 0's and *n* X's and ending up with X 0 X 0 ... X 0 after exactly *n* moves. It is a little complicated to describe, but the general strategy is systematically to work from each end to build up a 0 X X 0 0 X X ... type configuration.

The process starts as follows:

```
0 0 0 0 0 0 0 ................. X X X X X X X
0       0 0 0 0 0 ................. X X X X X X X 0 0
0 X X 0 0 0 0 0 ................. X X X     X X 0 0
0 X X 0 0       0 ................. X X X 0 0 X X 0 0
```

Continue in this vein, moving alternately pairs of 0's to the right and pairs of X's to the left. We have to separate cases according to the remainder when we divide *n* by 4. Suppose that $n = 4m$, a multiple of 4. (Try this with $n = 8$ and $n = 12$ to see what is going on.) After $2m - 2$ moves, we have gone from the starting configuration:

... O O O O O O O O X X X X X X X X ...

to the configuration:

... O X X O O O O O X X X X   X X ... > move $2m-1$
... O X X O O   O X X X X O O X X ... > move $2m$
... O X X O O X X O   X X O O X X ... > move $2m+1$
... O X X O O X X O X O X   O X X ... > move $2m+1$
... O X X O   X O X O X O X O X X ... > move $2m+2$

and now there are $2m-2$ remaining moves which will be described below.

Suppose that $n = 4m+1$, one more than a multiple of 4 (for example $n = 9$ or 13). After $2m-2$ moves, we have gone from the starting configuration to:

... O X X O O O O O O X X X X X   X X ... > move $2m-1$
... O X X O O   O O X X X X X O O X X ... > move $2m$
... O X X O O X X O O X X   X O O X X ... > move $2m+1$
... O X X O O X X O   X O X X O O X X ... > move $2m+2$
... O X X O O X X O X O X O X   O X X ... > move $2m+2$
... O X X O   X O X O X O X O X O X X ... > move $2m+3$

and there are $2m-2$ moves to go.

Suppose that $n = 4m+2$ (for example, $n = 10$ or 14). After $2m-2$ moves, the centre of the row looks like:

... O X X O O O O O O O X X X X X X   X X ... > move $2m-1$
... O X X O O   O O O X X X X X X O O X X ... > move $2m$
... O X X O O X X O O O X   X X X O O X X ... > move $2m+1$
... O X X O O X X   O X O O X X X O O X X ... > move $2m+2$
... O X X O O X X O X O X O   X X O O X X ... > move $2m+3$
... O X X O O X X O X O X O X O X   O X X ... > move $2m+3$
... O X X O   X O X O X O X O X O X O X X ... > move $2m+4$

and now 2m-2 moves remain.

Finally, suppose that n = 4m+3 (for example, n = 11 or 15). After 2m-2 moves, the centre of the row looks like:

```
... 0 X X 0 0 0 0 0 0 0 0 X X X X X X X     X X ... > move 2m-1
... 0 X X 0 0      0 0 0 0 X X X X X X X 0 0 X X ... > move 2m
... 0 X X 0 0 X X 0 0 0 0 X X X     X X 0 0 X X ... > move 2m+1
... 0 X X 0 0 X X 0     0 X X X 0 0 X X 0 0 X X ... > move 2m+2
... 0 X X 0 0 X X 0 X 0 0 X X     0 X X 0 0 X X ... > move 2m+3
... 0 X X 0 0 X X 0 X 0     X 0 X 0 X X 0 0 X X ... > move 2m+4
... 0 X X 0 0 X X 0 X 0 X 0 X 0 X 0 X     0 X X ... > move 2m+5
... 0 X X 0     X 0 X 0 X 0 X 0 X 0 X 0 X 0 X X ... > move 2m+5
```

and again $2m-2$ moves remain.

In all cases, with $2m-2$ moves to go, the centre of the row is now in the desired form. We now move alternately pairs X 0 to the left and 0 X to the right working outwards to the end until the task is complete.(M.C.)

FOUR PROBLEMS WITH WHOLE NUMBERS

1. Find all the whole numbers for which the number and its square together consist of exactly nine digits with each of the digits 1, 2, 3, ..., 9 appearing exactly once. (0 does not appear.)

2. You are given three numbers, each containing the distinct digits a, b, c in some order, namely (abc), (bca), and (cab), where

$$(abc) = 100a + 10b + c.$$

The product recorded for the three numbers is

$$234532286,$$

but unfortunately, this is not correct. All the digits are correctly given, but with the exception of the 6, all of them are in the wrong positions. Find the three numbers.

3. Find all positive integers with the property that when the cube and fourth power are written side by side, each of the ten digits 0, 1, 2, 3, ... ,9 appears exactly once.

4. Find all positive integers whose square is equal to the fifth power of the sum of its digits.

Hints

All of these problems require a bit of searching, so that it would be helpful to do a bit of preliminary reasoning to cut down the field a bit. One useful tool is that of casting out nines (see Appendix). For questions 1 and 3, you should first determine how many digits each of the numbers concerned has.

Solutions

1. Let N be the required number. The only way for N and N^2 to involve nine digits together is for N to have three digits and N^2 six. Since the smallest six digit number is 123456, $N \geq 352$. If N^2 begins with a 9, then so must N. Hence $N^2 \leq 876543$, so $N \leq 936$. N cannot end in 1, 5, 6, for, otherwise, N^2 would end in the same digit. Since any number is divisible by 9 if and only if its digital sum is 9, we see that $N(N+1) = N^2 + N$ is divisible by 9, so that either N or $N+1$ is a multiple of 9. Using these facts and some trial and error, we find that there are two possibilities:

$$567^2 = 321489$$

$$854^2 = 729316$$

2. The product of a, b, and c must be a number ending in 6. Therefore, none of a, b, c can be 0 or 5. None of the digits a, b, c can be 1 or 2. Suppose, for example, that a is 1 or 2. Since the larger of b and c does not exceed 9 and the smaller does not exceed 8, the product

$(abc)(bca)(cab) < (300)(1000)(900) = 270000000.$

But this is impossible, since the first digit of the product cannot be 1 or 2. Hence, each of the digits is at least 3. The possibilities for a, b, c in ascending order are 3, 4, 8 or 3, 6, 7, or 3, 8, 9, or 4, 6, 9, or 6, 7, 8. By casting out nines, we find that the product leaves a reminder 8 when divided by 9. Since (abc), (bca), (cab) all have the same digital sum, each must leave the same remainder when divided by 9. This eliminates all but the possibility 3, 8, 9. The number 389 does not work, but we find that

$$(398)\,(983)\,(839) = 328245326$$

3. Let n be the number we are looking for. If n does not exceed 9, then n^3 has no more than three digits and n^4 no more than four digits, so there cannot be 10 digits involved in the two powers. On the other hand, if $n \geq 22$, then n^3 needs at least five and n^4 six digits, which are too many. Thus, n^3 must have four digits and n^4 six digits. This narrows the possibilities down to 18, 19, 20, 21. By casting out nines, we find that $n^3 + n^4 = n^3(n + 1)$ must be divisible by 9, which eliminates $n = 19, 20$. But $n = 21$ is out because all its powers end in the same digit, 1. Thus, if there is an answer, it must be 18. Checking, we find that $n^3 = 5832$ and $n^4 = 104976$.

4. An obvious answer is 1. Are there any other answers? Let n be a possible answer and let s be the sum of its digits. The number $n^2 = s^5$ is at the same time a square and a fifth power. The only way this can happen is for it to be a tenth power, so that n is the fifth power of some number and s is the square of another number. Hence, the only possibilities for n are 1, $32=2^5$, $243=3^5$, $1024=4^5$, $3125=5^5$, $7776=6^5$, ... Checking these, we find that 243 works. $243^2 = 3^{10} = 9^5 = (2 + 4 + 3)^5$.

To see whether this exhausts all the possibilities, we need a more general analysis. Suppose that n has exactly k digits. Then

$$n \geq 10^{k-1} \text{ and } n^2 \geq 10^{2(k-1)}.$$

On the other hand, the sum s of the digits cannot exceed $9k$, so that

$$s^5 \le 9^5 k^5 < 10^5 k^5.$$

If $k \ge 6$, then

$$10^{2k-7} \ge k^5,$$

and we find that

$$n^2 \ge 10^{2(k-1)} > 10^5 k^5,$$

so that n^2 cannot be equal to s^5. Thus, all possibilities for n cannot have more than 5 digits.

Let d be the digital sum of n and, hence, of s. By casting out 9s, we see that d^2 and d^5 have the same digital sum, so that d must be 3, 4, 6, 9. Hence, the possible values of n reduce to numbers leaving these remainders upon division by 9. We have a finite number of numbers to check and we find that a complete collection of such numbers consists of 1, and 243.

TRIANGULAR ARRAYS OF DOTS

This diagram illustrates a triangular array of dots, seven to a side, for which all but one of the dots has been covered by disjoint equilateral triangles each covering three of the dots.

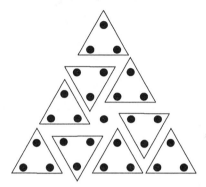

If we replace seven by some other number, we may find equilateral triangular arrays of dots with that number of dots along each side, *all* of whose dots can be covered by disjoint equilateral triangles each covering three of the dots. A trivial example is the array with two dots to a side which is included in one triangle covering three of the dots.

What other possible numbers of dots can a triangular array have for complete coverage by the smaller equilateral triangles with three dots?

Some Guidelines

Suppose that the array given has *n* dots to a side. Make a list with *n* in the first column and the number of dots in the array in the second. Are there any values of *n* for which the partition is clearly impossible? How do you know?

We know that the job can be done for $n = 2$. What is the next highest value of n for which a solution might be possible? Is there any way you can generate from a solution for a specific value of n a solution for a higher value of n?

We get to a higher value of n by adjoining extra rows of dots along the bottom of a triangular array. Therefore, one might look at the possibility of covering two or more adjacent rows of dots with small triangles.

The corner vertices of a triangular array can be covered by a small triangle in only one way. What can now be said about the cases $n = 3, 4, 5$?

Solution

The array with 7 dots to a side has 28 dots. The array with 9 dots to a side has 45 dots. There is a general formula for the number of dots in an array with n dots to a side; it is $\dfrac{n(n+1)}{2}$. For this number of dots to be covered by three-dot triangles, it must be divisible by 3. Thus, either n or $n+1$ must be a multiple of 3. The task is therefore impossible for

$$n = 1, 4, 7, 10, 13, 16, 19, \ldots.$$

The reader should be able to check without much difficulty that the task is impossible if $n = 3, 5, 6$. This leaves $n = 8$ as the smallest value of n exceeding 2 for which we might expect success. With a little more effort, we find that this too leads to failure.

However, when $n = 9$ and $n = 11$, we do get solutions:

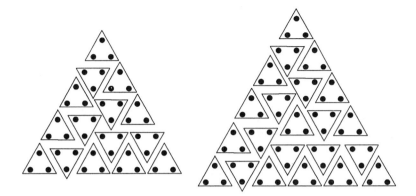

Having these in hand, we can get success for high values of n by building on solutions we already know.

There are two types of bands that can be partitioned into smaller triangles:

(1) Two rows of dots with $3k+1$ dots in the top row and $3k+2$ dots in the bottom (e.g. 7 and 8 respectively, 10 and 11 respectively)

(2) Three rows of dots with $2k-2$ dots in the top row and $2k$ dots in the bottom row (so that the number of dots in either row is even)

Using these two rules, we can list values of n for which we can expect success. The arrow between a smaller and larger value of n indicates that a solution for the smaller value generates a solution for the larger. You might want to check one or two cases yourself to see how the $n = 14$ cases are solved.

$$9 \rightarrow 11 \rightarrow 14$$
$$9 \rightarrow 12 \rightarrow 14$$

There is a more subtle way of getting new solutions from old. The diagram illustrates how to do it. The inner triangle has n dots to a side and the width of the bands is 3, so that the whole array including the inner triangle and bands has $n+9$ dots to a side. Suppose that n is even and we are given a solution for the inner triangle; then using rule (2) above for bands of width 3, we can get a solution for the whole array. The solution for $n = 11$ illustrates how we get a solution from the $n = 2$ case.

Now we are in a position to generate lots of values of n for which the problem can be solved:

$$12 \rightarrow 21 \rightarrow 24 \rightarrow 33 \rightarrow 36 \rightarrow 45 \rightarrow 48 \rightarrow \cdots$$

$$12 \rightarrow 14 \rightarrow 23 \rightarrow 26 \rightarrow 35 \rightarrow 38 \rightarrow 47 \rightarrow \cdots$$

Let us take stock:

We know that there are no solutions for these values of n:

$$1, 3, 4, 5, 6, 7, 8, 10, 13,$$
$$16, 19, 22, 25, 28, 31, \ldots$$

and that there are solutions for these values of n:

$$2, 9, 11, 12, 14, 21, 23, 24, 26,$$
$$33, 35, 36, 38, 45, 47, 48,$$

How about the remaining values of n:

$$15, 17, 18, 20, \ldots?$$

This, dear reader, is for you to decide.

THE ROTATING TABLE

The Russians have a reputation for creating beautiful problems. Here is one I found out about a decade or so ago from Claude Gaulin at Laval University.

Imagine a square table, the top of which can rotate freely on a central pedestal. In each corner of the table is a deep well containing a tumbler that cannot be seen and is either upright or inverted. Your task is to put all the tumblers the same way (either all upright or all inverted) in a finite number of moves.

For each move, the table rotates and then stops at random. You are permitted to feel two of the tumblers and turn over as many as you wish—both, one or none. If by chance, all tumblers become uniform, a bell sounds and you know the job is done. Otherwise, the table rotates and another move is made.

Can you guarantee that, after a finite number of moves, the task is completed? If so, how?

You might think not. After all, it is conceivable that some tumbler might always escape attention. But note that it is not specified which way up all the tumblers have to be, just so long as they are the same way up. So, go ahead and solve the problem.

. . . .

If you want a greater challenge, solve the problem under the more stringent requirement that you are not allowed to reach into the wells. Rather, all you can do is to instruct a robot to reach into two of the wells as indicated by you, and flip one or both of the tumblers.

The Solution

If you are allowed to reach into the wells, then at most five moves are required. At each stage, you have two choices: pick a diagonal pair of wells or an adjacent pair. After making a diagonal and an adjacent move in either order, you can ensure either that the bell has sounded or that you are left with three up and one down (UUUD in cyclic order). Feel a diagonal pair; if one is down, flip it over and the bell will sound; otherwise, flip one over to get UUDD. Feel an adjacent pair; flip both to make the bell sound or get UDUD. Finally, flip a diagonal pair to complete the task.

. . . .

For the version with the robot, you may need seven moves. There are four possible states:

(1) three one way up, one the other way;

(2) an adjacent pair one way, and an adjacent pair the other way;

(3) a diagonal pair one way, and a diagonal pair the other way;

(4) all the same way, with the bell sounding.

Here are the moves; if (4) occurs you are done; if (4) does not occur, go to the next move.

- flip a diagonal pair to get (1), (2) or (4);
- flip an adjacent pair to get (1), (3), or (4);
- flip a diagonal pair to get (1) or (4);
- flip one glass to get (2), (3), or (4);
- flip a diagonal pair to get (2) or (4);
- flip an adjacent pair to get (3) or (4);
- flip a diagonal pair to get (4).

PROBLEMS FOR AN EQUAL-ARM BALANCE

1. You are given six pucks, two coloured red, two blue, and two white. They are identical in appearance. One puck of each colour has the same lighter weight; the other puck of each colour has the same heavier weight.

   Using an equal-arm balance twice, determine the heavier and the lighter of each type.

2. You have five billiard balls, identical in appearance. Three of them weigh exactly the same; one of the remaining balls is heavier than the rest; the other is lighter. The heavy and light balls together exactly balance two of the others. Using an equal-arm balance the least number of times, determine the heavy and the light balls.

3. You have nine billiard balls, identical in appearance. Seven of them weigh the same, and the other two—one heavy and one light—exactly balance any two of the seven equal balls. Using an equal-arm balance the least number of times, determine the heavy and the light balls.

Comments

To analyze this type of problem, you might consider how many possibilities there are for the answer. For example, in problem 2, there are five possibilities for the heavy ball, and for each of these, four possibilities for the light ball. Thus, there are $5 \times 4 = 20$ possibilities for the answer. Each use of the equal-arm balance has these possible outcomes: the pans balance (B), the left pan descends (L), and the right pan descends (R). Each of the 20 possibilities for the heavy-light pair is consistent with precisely one of the balance outcomes. Knowing the result of the first use of the balance will narrow us down to a subset of the 20 possibilities.

The effectiveness of our procedure can be measured by the largest number of possibilities that might be outstanding at the end of each use of the balance. Since any sum of three positive integers that add up to 20 must contain one number at least equal to 7, one of the balance outcomes will leave us with at least 7 possibilities to mediate among. Can you bring this worst case number as close to 7 as possible?

Using this analysis, you can first get a lower bound for the number of balances that you would need to cover all eventualities.

Solutions

1. Denote the pucks by R, r, B, b, W, w, where the letters indicate colours. First compare R, W with r, B. If the two sides balance, then R and B are light while r and W are heavy, or else R and B are heavy while r and W are light. A comparison of two balls of the same colour will now settle the matter.

   If, say, R and W is heavier than r and B, then R is heavier than r and we cannot have that both W is light and B is heavy. Now compare W, B with w, b to settle the matter.

2. There are many ways to solve this problem, and your solution may be perfectly correct. Here is one way to approach the situation:

   Let us suppose the balls to be numbered

   ❶ ❷ ❸ ❹ ❺

   There are twenty possibilities for the heavy-light pair, and we have to find the correct one. The best we can do after one use of the balance is to leave 7 outstanding possibilities. Of these at least seven possibilities, at least 3 will be outstanding after the second use of the balance. Therefore, we will need to use the balance no fewer than three times.

Suppose we weigh two balls against two. If the pans balance, there are four possibilities—one of the two pairs on the pans being the light and the heavy balls. If the left pan descends, then there are eight possibilities: one of the left pan balls is heavy and one of the right pan balls is light (4 possibilities), one of the left pan balls is heavy and the unused ball is light (2 possibilities), or one of the right pan balls is light and the unused ball heavy (2 possibilities). The descent of the right pan also corresponds to eight possibilities.

This regime of balancings will solve the problem:

- First test: balance ❶ ❷ against ❸ ❹
- Second test: balance ❶ ❸ against ❷ ❹
- Third test: balance ❶ ❹ against ❷ ❸

The outcomes BLL for example inform us that ball ❶ is heavy and ball ❷ is light. The twelve possibilities for which the odd balls are among ❶ ❷ ❸ ❹ corresponds to the twelve outcomes in which the pans balance exactly once.

The eight possibilities for which ❺ is one of the odd balls correspond to the eight outcomes in which the pans never balance.

Regardless of which two balls are odd, it is not possible for the pans to balance two or three times.

2. A somewhat similar strategy will lead to a solution when there are nine balls. There are 72 possible configurations, so we try to arrange for 24 of the possibilities to correspond to each outcome of the first use of the balance. After the second use, we would like to reduce the ambiguity to 8 possibilities.

Here is one way to do it.

- First test: balance ❶ ❷ ❸ ❹ against ❺ ❻ ❼ ❽
- Second test: balance ❶ ❷ ❺ ❻ against ❸ ❹ ❼ ❽
- Third test: balance ❶ ❸ ❺ against ❷ ❹ ❻
- Fourth test: depends on what we learn from the first three

In the table below, we indicate the results of the use of the balance three times, the numbers of the ball in the odd pair with the heavier ball listed first that are consistent with the outcome, and the balls to use in the final balance.

BBB	❼❽, ❽❼	❼ against ❽
BBL	❶❷, ❸❹, ❺❻	❶ against ❸
BBR	❷❶, ❹❸, ❻❺	❶ against ❸
BLB	❶❸, ❷❹	❶ against ❸
BLL	❶❹, ❺❼, ❺❽	❼ against ❽
BLR	❷❸, ❻❼, ❻❽	❼ against ❽
BRB	❸❶, ❹❷	❶ against ❸
BRL	❸❷, ❼❻, ❽❻	❼ against ❽
BRR	❹❶, ❼❺, ❽❺	❼ against ❽
LBB	❶❺, ❷❻	❶ against ❸
LBL	❶❻, ❸❼, ❸❽	❼ against ❽
LBR	❷❻, ❹❼, ❹❽	❼ against ❽
LLB	❾❼, ❾❽	❼ against ❽
LLL	❶❼, ❶❽, ❶❾	❼ against ❽
LLR	❷❼, ❷❽, ❷❾	❼ against ❽

LRB	35, 46	1 against 3
LRL	36, 39, 96	3 6 against 1 2
LRR	46, 49, 96	4 6 against 1 2
RBB	51, 62	1 against 3
RBL	52, 74, 84	7 against 8
RBR	61, 73, 83	7 against 8
RLB	53, 64	1 against 3
RLL	54, 94, 59	4 6 against 1 2
RLR	63, 93, 69	3 6 against 1 2
RRB	79, 89	7 against 8
RRL	72, 82, 92	7 against 8
RRR	70, 80, 90	7 against 8

ISOMERS

Hydrocarbons are chemical compounds made up of carbon (C) and hydrogen (H) atoms, each linked to its neighbours by bonds. A hydrogen atom has one bond linking it to a single adjacent carbon atom. A carbon atom has exactly four bonds which link it to up to four adjacent atoms. Single, double and triple bonds can exist between a pair of carbon atoms. Since each molecule has to hang together, each pair of atoms is joined by a bonded chain.

Here are two examples of hydrocarbons:

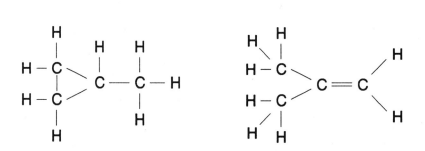

These illustrate that the configuration is not uniquely determined by specifying the number of atoms of each type. Both diagrams answer to the formula C_4H_8. In all there are five distinct structures with this formula. Chemical compounds having distinct molecular structures but the same formula are called isomers. Can you find the other three? (Structures are distinct if one cannot be twisted around to form another while keeping all the bonds intact.)

Design all realizations of the formula C_6H_{12} and C_8H_{18}.

Comments

We can abbreviate the process of finding and recording solutions by keeping in mind a few basic considerations. Since a hydrogen atom cannot be joined to two carbon atoms or to any other hydrogen atom, the carbon atoms must be bonded together. In making a diagram, we need not insert the hydrogen atoms, since there is only one way to complete a diagram with hydrogen atoms once the carbon array is given.

Suppose that the formula we are dealing with is $C_m H_n$. Let us find the total number of bonds. There are four bonds for each of the m carbon atoms and one for each hydrogen atom, giving a count of $4m + n$. But this counts each bond twice, since it joins two atoms. Therefore the actual number of bonds must be $\frac{4m + n}{2}$. Since n of these bonds involve a hydrogen atom, there must be $\frac{4m + n}{2} - n = \frac{4m - n}{2}$ bonds joining two carbon atoms.

Thus, for $C_4 H_8$ there must be 4 bonds joining carbon atoms. Argue that there must either be exactly one double bond or one loop and no double bonds. For $C_6 H_{12}$, we must have 6 bonds joining carbon atoms, while for $C_8 H_{18}$, we must have 7 bonds.

One way to sort out the possibilities is to do it according to the number of bonds of each of the m carbon atoms; this amounts to partitioning the number $4m - n$ (twice the number of bonds since each is counted twice) into m parts. Thus, for $C_4 H_8$, the first example in the statement of the problem exemplifies the partition $8 = 3+2+2+1$, since there is one carbon atom with three bonds to other carbon atoms, two carbon atoms with two bonds to other carbon atoms and one carbon atom with only one bond to another carbon atom. The second example exemplifies the partition $8 = 4+2+1+1$. (F.H.R.)

The Solution

According to this classification scheme, we list the formula alone with the partitions of $4m - n$ and the number of distinct possibilities.

C_4H_8		C_6H_{12}		C_8H_{18}	
(3 2 2 1)	2	(4 4 1 1 1 1)	1	(4 4 1 1 1 1 1 1)	1
(4 2 1 1)	1	(4 3 2 1 1 1)	5	(4 3 2 1 1 1 1 1)	3
(2 2 2 2)	1	(4 2 2 2 1 1)	4	(4 2 2 2 1 1 1 1)	3
(3 3 1 1)	1	(3 3 3 1 1 1)	2	(3 3 3 1 1 1 1 1)	1
		(3 3 2 2 1 1)	8	(3 3 2 2 1 1 1 1)	5
		(3 2 2 2 2 1)	4	(3 2 2 2 2 1 1 1)	4
		(2 2 2 2 2 2)	1	(2 2 2 2 2 2 1 1)	1
Total	5	Total	25	Total	18

For C_4H_8, the diagrams are

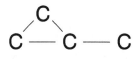

methylcyclopropane

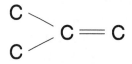

methylpropene

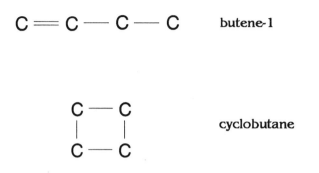

C ═ C — C — C   butene-1

C — C
C — C   cyclobutane

For the fifth possibility, while there is only one way of doing it mathematically, there are actually two distinct chemical substances involved:

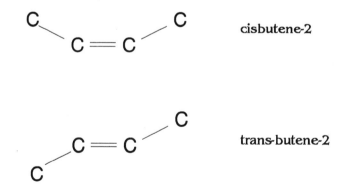

cisbutene-2

trans-butene-2

While there can be free rotation around a single bond, there cannot be around a double bond. So the cis- and trans- isomers are chemically distinct. For example, the trans- version has higher boiling and melting points than the cis- version. [A.C. & D.S.]

Here are the five versions of C_6H_{12} corresponding to the partition (4 3 2 1 1 1). The reader will be left with the task of tracking down the remaining isomers.

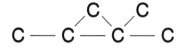

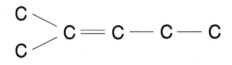

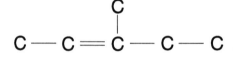

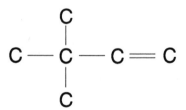

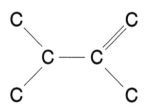

TWO DIOPHANTINE EQUATIONS

Diophantus was an Alexandrian mathematician who lived sometime during the first three centuries of the Christian Era. Because his *Arithmetic* consisted of thirteen volumes devoted in great part to numerical equations, his name is given to algebraic equations for which solutions in integers—positive, negative or zero —are sought.

For example, the diophantine equation

$$x^2 + xy - y^2 = 1$$

is satisfied by $(x, y) = (-2, 1)$, $(1, 0)$ and $(2, 3)$. Can you find other solutions to this equation? In fact, you can keep producing new solutions as long as you want.

Here is a more formidable-looking diophantine equation:

$$(x^3 + y^3 + z^3) + (xz^2 - xy^2 - yz^2) + (2x^2z - 3xyz) = 1$$

You may wish to check that this is satisfied by

$$(x, y, z) = (2, 4, 3), (1, 1, -1) \text{ and } (2, 0, -1).$$

There are lots more solutions, as well as a nifty little rule that lets you generate new solutions from known ones.

See what you can find out.

Comments

You should probably begin with a little trial and error. Get as many numerical solutions in front of you as you can—they are not hard to find. Then begin to look for patterns. Look in particular at solutions of the second equation where x, y, z are all positive, and put them in a list in increasing order of magnitude. There is an experimental side to mathematics. One often just assembles a lot of data, and then makes conjectures about what might be true in general. In this case, you will find that, once you get on the right track, you can keep generating solutions until the cows come home.

Solution

$$x^2 + xy - y^2 = 1$$

The Fibonacci sequence 0, 1, 1, 2, 3, 5, 8, 13, 21, 34, 55, ... with each term the sum of its two predecessors) is tied up with this equation. Here are some solutions:

$$(x, y) = (1, 1), (2, 3), (5, 8), (13, 21), (34, 55), ...,$$

$$= (-1, -1), (-2, -3), (-5, -8), ...$$

$$= (1, 0), (2, -1), (5, -3), (13, -8), (34, -21), ...$$

$$= (-1, 0), (-2, 1), (-5, 3), (-13, 8), ...$$

If we let $f(x, y) = x^2 + xy - y^2$, then it turns out that

$$f(u, v) = f(u + v, u + 2v) = f(2u - v, v - u).$$

Beginning with a known solution $(x, y) = (u, v)$, we can generate a chain by taking next $(x, y) = (u + v, u + 2v)$ or $(2u - v, v - u)$.

Another approach is to start with a solution $(x, y) = (u, v)$.

Consider the following quadratic equation in x:

$$x^2 + vx - (v^2 + 1) = 0.$$

This equation has integer coefficients and an integer solution $x = u$. The theory of the quadratic tells us that the sum of its solutions is the coefficient $-v$. Thus, its second solution is $x = -u - v$. Therefore, the solution $(x, y) = (u, v)$ gives rise to a new solution $(x, y) = (-u-v, v)$.

Similarly, by considering the quadratic equation in y,

$$y^2 - uy - (u^2 - 1) = 0,$$

we find that the solution $(x, y) = (u, v)$ generates the solution $(u, u-v)$. Starting with the obvious $(x, y) = (1, 0)$ and using these two ideas, we are led to the chains

$$(1, 0), (-1, 0), (-1, -1), (2, -1), (2, 3), (-5, 3), (-5, -8),$$

$$\cdots$$

and

$$(1, 0), (1, 1), (-2, 1), (-2, -3), (5, -3), (5, 8), \ldots$$

$$\cdots \cdots$$

$$(x^3+y^3+z^3) + (xz^2-xy^2-yz^2) + (2x^2z-3xyz) = 1$$

Before giving a rule to generate some solutions from others, let us list a few solutions:

(1, 0, 0)	(1, 0, 0)
(0, 1, 0)	(-1, 0, 1)
(0, 0, 1)	(1, 1, -1)
(1, 1, 0)	(0, -1, 1)
(0, 1, 1)	(-1, 1, 0)
(1, 1, 1)	(2, 0, -1)
(1, 2, 1)	(-2, -1, 2)
(1, 2, 2)	(1, 2, -2)
(2, 3, 2)	(1, -2, 1)

(2, 4, 3)	(–3, 1, 1)
(3, 5, 4)	(4, 1, –3)
(4, 7, 5)	(–3, –3, 4)
(5, 9, 7)	(0, 4, –3)
(7, 12, 9)	(4, –3, 0)
(9, 16, 12)	(–7, 0, 4)
(12, 21, 16)	(7, 4, –7)

How many of these solutions did you find?

If we take any two consecutive solutions in the first column and add corresponding entries, we get the second following solution. For example,

$$(2, 3, 2) + (2, 4, 3) = (4, 7, 5).$$

If we take any two consecutive solutions in the second column and add corresponding entries, we get the second previous solution. For example,

$$(4, 1, –3) + (–3, –3, 4) = (1, –2, 1).$$

Try to extend the two columns using these ideas, and check that you receive new solutions.

In a different vein, if we let

$$g(x,y,z) = x^3 + y^3 + z^3 + xz^2 - xy^2 - yz^2 + 2x^2z - 3xyz,$$

it turns out that

$$g(u, v, w) = g(w, u+w, v) = g(v-u, w, u).$$

Thus, if $(x, y, z) = (u, v, w)$ satisfies the equation, then so also do $(x, y, z) = (w, u+w, v)$ and $(v-u, w, u)$. The mapping $(u, v, w) \rightarrow (w, u+w, v)$ takes us down the first column in the above list, while $(u, v, w) \rightarrow (v-u, w, u)$ takes us down the second.

· · · · · ·

Those looking for a harder challenge can try to generate some solutions of the equation $x^3 + 2y^3 + 4z^3 - 6xyz = 1$ in whole numbers.

A MAGIC HEXAGON

The square array of numbers

$$4\ 3\ 8$$
$$9\ 5\ 1$$
$$2\ 7\ 6$$

has its origin in ancient China. It is a "magic" square: the sum of the numbers in each of its rows, columns and diagonals is the same: 15. There is a vast literature on magic squares of different dimensions. However, we can ask for other sorts of magic polygons.

Place the numbers 1, 2, 3, ..., 17, 18, 19 in the cells of the array at the right to form a magic hexagon in which the numbers in each row of adjacent cells (having a common side) parallel to a side of the hexagon have the same sum.

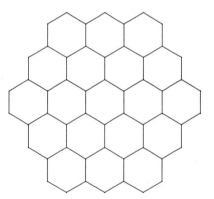

Discussion

This is a very difficult problem. Yet within two weeks of its appearance in the *University of Toronto Alumni Magazine*, John R. Bird of Toronto had submitted a solution. Six weeks later, a solution came in from Leslie J. Upton of Mississauga. This is a remarkable achievement since the solution turns out to be unique up to reflection and rotation of the figure. The story of this problem is told in Ross Honsberger, *Mathematical Gems* (Dolciani Mathematical Expositions #1), Mathematical Association of America, 1973.

The first thing to do is to determine the common sums of each row. What is the sum of the numbers from 1 to 19 inclusive? Note that the figure is made up of five disjoint rows in one of the directions.

Let us mark the positions by letters:

Consider the triangular array consisting of D, E, F, J, K, P. The remaining letters belong to four distinct rows, so that one can find $D + E + F + J + K + P$. But one also knows $D + E + F + G$.

Notice also that

$$(B + F + L + R) + (C + G + M)$$
$$= (T + Q + L + G) + (V + R + M).$$

Cancel terms common to both sides. One ultimately arrives at $F + S = A + Q = M + J$.

Also, show that $D + G + T = N + R + B = K +$ (row sum) and $N + Q = F + D$. (J.R.B.)

Solution

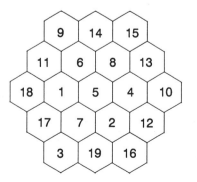

Chapter Five

Other Problems and College Tales

DOUBLE CHRISTMAS

The famous aviatrix, Emily Airborne, left Sydney, Australia, in her jet at 7 a.m. local time on December 26 and flew northeasterly to Toronto. After one hour of flight, she crossed her first time zone boundary, and then, at intervals of two hours exactly, crossed eight more time-zone boundaries. Except for the third of the nine, the International Dateline, where she had to retard her watch by 23 hours, she advanced her watch by one hour when she crossed each time-zone boundary. Assuming she rested for 36 hours before and after the trip, determine how many hours Emily experienced of each of Christmas Day and Boxing Day. (Christmas Day is December 25; Boxing Day, December 26.)

Hints

It does not really matter exactly what time she landed in Toronto, as she is on Toronto time once she passes the final time zone. Taking into account her travelling time and the time change, how much does her watch advance from having crossed one time zone to having crossed the next? When on her journey will she first experience Christmas Day; what time is it? It might be useful to draw a line diagram, indicating time zones and watch readings on either side of the time zone.

Solution

Emily had 30 hours of Christmas Day, 24 hours in Sydney and 6 hours en route, and 33 hours of Boxing Day, 7 hours in Sydney, 15 hours after the last time zone (including time on the ground in Toronto), and a total of 11 hours en route to the last time-zone boundary.

The International Date Line is the third time-zone boundary. At the instant after it is crossed, Emily has flown 5 hours and her watch reads 5 + 3 = 8 hours later than 7 a.m., namely 3 p.m. It is now 3 p.m. on December 25.

After crossing three more boundaries, she has flown an additional 6 hours and her watch has advanced 6 + 3 = 9 hours to midnight. It is once again December 26.

There remain three more time-zone boundaries to cross. As she crosses the last one, she has flown 6 hours and it is 9 a.m. At this point, she has had 6 hours of her second Boxing Day and will enjoy an additional 15 hours.

A MARRIAGE PROBLEM

The village leader poured the anthropologist another cup of the local brew as they warmed themselves by the evening fire. "We live in a very orderly society here," the leader intoned. "Everyone is married by the age of 18. In fact, I choose the brides for the young men myself."

"Don't you find," enquired the anthropologist, "that there is some resentment about this?"

"One just cannot do these things carelessly. I have found a system that works very well," continued the leader. "You have to avoid a situation in which a boy and a girl each prefer the other to the spouses to which I have assigned them. If, say, a young man prefers another woman to his own wife, it doesn't do him any good. That other woman would rather remain with her husband."

"And you manage to provide each young person with a mate?"

"Yes, certainly. We always have the same number of boys and girls when the choosing ceremony takes place. Each girl and each boy is asked to list in order of preference the members of the opposite sex. These lists are the basis of my selection."

"How do you do it?"

"I am afraid I will have to tell you in the morning. It is now time for me to retire to my hut to watch *Rumpole of the Bailey*." The anthropologist lay awake for a considerable part of the night trying to figure out the method. "Suppose," he thought, "we assign to each boy the girl at the top of his list. If all the girls were different, then that would do it: each boy would have the one he likes best. If however, two boys wanted the same girl, then she could pick the better one; the second boy would have to be assigned to another girl. Is there one available?" All these thoughts made his head spin. Can the reader help out?

* * * * * *

The next morning, over a breakfast of grasshopper pie, the anthropologist raised the subject again. "You got off to a good start, except that I always look at the preferences of the girls first rather than the boys. You can think of the process proceeding through a number of rounds.

"On the first round, each woman proposes to the man at the top of her list. Some men may receive more than one proposal, others none. If he receives a proposal, he agrees to hold the most favoured one for consideration while rejecting any others.

"On the second round, all the women who have been rejected on the first round propose to the next person on the list. The men hold for consideration the best proposal among the one from the first round and those of the second, and refuse all the rest.

"The rounds continue. At the beginning of each round, the women cross off their lists the names of all those who have given a refusal and propose to the best remaining man. The man keeps for consideration the best proposal he has received to that round."

"How do we know that this does not go on forever?" asked the anthropologist.

"If in any round, every woman has her proposal accepted, then the process stops, and I assign the couples accordingly. Every time the process continues, it is because at least one woman has had a refusal and has to cross a name off her list. Now there are only a finite number of names that can be crossed off, since there are only a finite number of women and men. So the process must terminate.

"When the process terminates, every man must hold a proposal. If not, there would be some man who has never received a proposal. Then, since the number of men receiving proposals is smaller than the number of women, there would have to be a woman who has been rejected by every man she proposed to. But this would force another round. Thus, at the end, each man has under consideration a proposal from a woman, and I make my assignment on that basis."

"But how do we know that no man and woman can defect from their chosen mates to be with each other?"

"Suppose I have assigned man x to woman X, and man y to woman Y. Suppose also that X prefers y to x. Then X must have proposed to y before x; y must have turned her down as he already had someone better under consideration. That someone better might not have been Y, but when Y eventually came along, she was clearly superior in y's eyes to anyone else seen so far. Therefore, we cannot have X preferring y while at the same time y prefers X."

Comments and Proposed Solution

This problem of course has been discussed in the literature. Here are some readable references:

> D. Gale & L.S. Shapley, "College admissions and the stability of marriage" *American Mathematical Monthly* 69 (1962), 9-15.

> Robert W. Irving, "Stable matching problems" *Mathematical Spectrum* 18 (1985/86), 6-14.

Many who responded to this problem suggested a variant of the following. For each pair consisting of a woman Z and a man z, associate a number pair (m,n), where Z rates z mth on her list and z rates Z nth on his. For example, if we have a joint rating $(1, 1)$, each most prefers the other and we marry them off. In general, for each possible pair, we form $m + n$; marry off the pairs with the smallest sum, delete from the list all such pairs involving either members of the married couples, and continue on.

However, here is a situation in which this process can run into trouble. Suppose the four women A, B, C, D and the four men a, b, c, d have the following preference lists:

$$a : A\,D\,C\,B \qquad A : d\,b\,a\,c$$
$$b : D\,C\,A\,B \qquad B : a\,b\,c\,d$$
$$c : C\,A\,D\,B \qquad C : c\,a\,d\,b$$
$$d : D\,C\,A\,B \qquad D : d\,c\,b\,a$$

Clearly, we can begin pairing c with C and d with D.

There are four pairs left to consider:

A and *a* with number pair (3, 1) and sum 4

A and *b* with number pair (2, 3) and sum 5

B and *a* with number pair (1, 4) and sum 5

B and *b* with number pair (2, 4) and sum 6

Thus, we should marry off A and *a*, leaving only B and *b* to be married. But A prefers *b* to *a*, and *b* prefers A to B, leaving the possibility of A and *b* indulging in hanky-panky.

BALLS IN A BAG

A bag contains 16 billiard balls, some black and the remainder white. Two balls are drawn at the same time. It is equally likely that the two balls will be the same colour as different colours. How are the balls divided within the bag?

Solution

If you said that there were eight balls of each colour, then you are wrong. There are 120 ways of choosing a pair of balls from the bag. With eight of each colour, 64 of these ways would yield one black and one white ball, while 56 ways would yield two balls of the same colour. (To see this, note that there are 8 ways to get a black ball and 8 ways to get a white. There are 28 possible pairs of black balls and 28 possible pairs of white balls. To get a mixed draw, for each of the 8 choices of white balls, there are 8 choices of black balls for a total of 64 choices for the pair.) If the drawing of the pair is random, each of the 120 pairs has an equal chance of being picked, so you would have a probability of $\frac{64}{120} = 0.53$ of getting one of each colour and a probability of $\frac{56}{120} = 0.47$ of getting of getting both of the same colour.

If, however, the bag contains 10 of one colour and 6 of the other, the there are 60 ways of drawing two balls of different colours, 45 ways of drawing two balls of the more prevalent colour, and 15 ways of drawing two balls of the less prevalent colour.

Comments

Suppose the ball contains x black balls and y white balls. What are the possible values of the positive integers x and y for which it is equally likely that two balls selected at random will be the same colour as different?

Before we answer this question, let us review the ways of counting how many ways we can choose a pair of elements from n elements. There are n ways of choosing one element, and, having chosen the first, $n-1$ ways of choosing the second. Hence, there are $n(n-1)$ ways of choosing the two elements. However, for each pair, this counts each order of choosing them once, so that each pair gets counted twice. So there are actually half this many distinct pairs, $\dfrac{n(n-1)}{2}$. For example, the number of choices of two letters from the five letters a, b, c, d, e is 10, namely ab, ac, ad, ae, bc, bd, be, cd, ce, de.

There are xy ways of choosing a black ball and a white ball. The number of ways of choosing two black balls is $\dfrac{x(x-1)}{2}$ and the number of ways of choosing two white balls is $\dfrac{y(y-1)}{2}$. The condition we want is that

$$xy = \frac{x(x-1)}{2} + \frac{y(y-1)}{2}.$$

This simplifies to $2xy = x^2 - x + y^2 - y$, or

$$(x - y)^2 = x + y.$$

We cannot have $x = y$, for then the left side would be zero and the right side nonzero. Let us suppose that x is greater than y, and suppose that $z = x - y$.

Then x and y satisfy the two equations

$$x - y = z \text{ and } x + y = z^2.$$

Hence

$$x = \frac{z(z+1)}{2} \text{ and } y = \frac{z(z-1)}{2}.$$

We can select for *z* any positive integer and get a value of *x* and *y* that works. Thus, the pairs (*x*, *y*) for which the two probabilities (drawing balls of different and same colours) are equal are:

(1, 0) (in this case, a draw of two balls is impossible)
(3, 1), (6, 3), (10, 6), (15, 10), (21, 15), ...

Observe that in every case, the number of balls in the bag is a perfect square.

PROTOCOL

The new janitor looked in amazement at the College dignitaries at high table. After they finished their meal, there was a flurry of standing and sitting until all were on their feet and they swept from the room. He asked the Bursar to explain.

"The tradition," she replied, "goes back to the earliest days. When the meal is over, the Principal can move (either stand or sit) at will. However, the Vice-Principal can move only while the Principal is seated. The Registrar can move only while the Vice-Principal is seated and the Principal is standing. The Dean of Men can move only while the Registrar is seated and the Vice-Principal and Principal are standing. Finally, the Dean of Women can move only while the Dean of Men is seated and the other three are standing.

"It seems complicated, but after a while they get quite good at mastering the moves and getting into the common room before the coffee gets cold."

Of course, when new officers are appointed to the College, it is necessary to teach them the routine. Draw up a procedure with the smallest number of moves that can be deposited in the College Archives.

Discussion

Since no one depends on the state of the Dean of Women to determine what he or she can do, this suggests that the first order of business is to get the Dean of Women on her feet. After this, we only need to get the other four standing. This suggests that the solutions for five people can be reduced to the solution for four, which in turn can be reduced to the solution for three and so on.

Thus, you might try to solve the problem when only the Principal and Vice Principal are present. Then, add the Registrar, and finally the Deans. At this stage, you should be able to see a pattern and extend the problems to any number of people with a similar hierarchical ordering.

Solution

Here is the correct answer. The letters refer to those standing at the end of each move; the meaning of the letters should be clear.

P				
P		R		
		R		
	V	R		
P	V	R		
P	V	R		W
	V	R		W
		R		W
P		R		W
P				W
				W
	V			W
P	V			W
P	V		M	W
	V		M	W
			M	W
P			M	W
P		R	M	W
		R	M	W
	V	R	M	W
P	V	R	M	W

George Baker of Kentville, Nova Scotia, puts the protocol as follows:

Now, if you like your coffee hot, hear this. You, Mr. Principal, make the first move and every second move thereafter until all are standing. Mr. Vice: you make the fourth move and every fourth thereafter. Eight, twelve, sixteen, and twenty to be exact. Mr. Registrar: you make the second move and every eighth thereafter. Remember: the second, tenth, and eighteenth. The Dean of Women will rise at the sixth move and the Dean of Men on the fourteenth. Yours not to reason why...

Another approach is to assign each person a number (P = 1, V = 2, R = 3, M = 4, W = 5), and record the solution as a sequence of numbers to indicate who stands or sits on the successive moves. (G.H.) Thus, the solution is:

$$1\ 3\ 1\ 2\ 1\ 5\ 1\ 2\ 1\ 3\ 1\ 2\ 1\ 4\ 1\ 2\ 1\ 3\ 1\ 2\ 1$$

Consider the generalization of the problem to n people, ranked in order. All are initially seated; the first ranked can move at will; each of the others can move only if the next immediately higher-ranked is seated and the other higher-ranked are standing. The object is to get everyone standing.

Denote by U_n the number of moves necessary to get everyone on their feet. We have the following table:

n	U_n	Solution
1	1	1
2	2	2 1
3	5	1 3 1 2 1
4	10	2 1 4 1 2 1 3 1 2 1
5	21	1 3 1 2 1 5 1 2 1 3 1 2 1 4 1 2 1 3 1 2 1
6	42	2 1 4 1 2 1 3 1 2 1 6 1 2 1 3 1 2 1 4 1 2 (then as n = 5).

Observe that $U_n = 2U_{n-1}$ when n is even and $U_n = 2U_{n-1} + 1$ when n is odd. The form of the values of U_n is striking when written in base 2:

$$1, 10, 101, 1010, 10101, 101010, \ldots$$

Let us formulate a related problem, in which n people are initially standing and we want to get them all seated following exactly the same rules. Let D_n be the number of moves necessary to do this. We have:

n	D_n	Solution
1	1	1
2	2	1 2
3	5	1 2 1 3 1
4	10	1 2 1 3 1 2 1 4 1 2
5	21	1 2 1 3 1 2 1 4 1 2 1 3 1 2 1 5 1 2 1 3 1

How are U_n and D_n related? Why?

RECYCLE BOOKS

Each year, the College held a huge sale of used books to raise money. As you can imagine, there were many faithful supporters who came year after year. Indeed, the Principal remarked, "I have seen Mr. Au, Mr. Baj, Miss Cook, and Mrs. Dan at three consecutive booksales."

"Let me tell you something remarkable," added the Alumni Director. "In 1993, each donated one book to the sale and purchased a book which another of the four had donated."

"And," commented the janitor, "in 1994, each donated that book to the sale and bought one of the four he or she did not own before."

"I suppose that now, in 1995, each has again donated the book bought last year and purchased the only one not previously owned."

"Right on," declared the Alumni Director. "I have kept track of the books and can tell you that Mr. Au is the only one to own *Moby Dick* before owning *Youth*."

"Well," remarked the Janitor, "I know that the same number of people owned *Youth* before *Mutiny On the Bounty* as owned it after."

"It makes you wonder," joshed the Principal, "who bought *The Old Man and the Sea* in 1994."

Who did buy the book?

Comments

When the problem originally appeared, there were several more clues, sufficiently many to determine a complete list as to who bought what when. One of the solvers, Michael Treadwell, a professor of English Literature at Trent University, drew my attention to the redundancy. If we add the clue that Miss Cook is the only one to own *Moby Dick* after owning the title that Mrs. Dan ended up with, then the whole situation becomes clear.

Most solvers started with a table with columns headed by names and rows headed by dates. Treadwell's approach was to try to get the configuration of books held by four people over the years, and then match the names in afterwards.

Solution

Mr. Au is the only one to own *Moby Dick* (*MD*) before owning *Youth* (*Y*). Consequently, he must have had *MD* at the start and *Y* at the end, as no one else could have had either possibility without contradicting the condition. The one who had *Y* in 1994 must have had *MD* in 1995 (the final year); the one who had *Y* in 1993 must therefore have had *MD* in 1994; finally the original owner of *Y* must have had *MD* in 1993. Thus, denoting the unknown people by 1, 2, 3, we can make this table:

	Mr. Au	1	2	3
Owned at start	*MD*	*Y*		
Bought in 1993		*MD*	*Y*	
Bought in 1994			*MD*	*Y*
Bought in 1995	*Y*			*MD*

Since the same number owned *Y* before *Mutiny on the Bounty* (*MB*) as after, Mr. Au and person 3 must have owned *MB* before *Y*. Hence person 2 must have owned *MB* last. Therefore person 1 bought *MB* in 1994, so that Mr. Au must have bought it in 1993. Therefore, Mr. Au bought *The Old Man and the Sea* (*OMS*) in 1994.

Let us now bring in the assumption that Miss Cook is the only one to own *MD* after owning the title that Mrs. Dan ended up with. We know that Mrs. Dan did not end up with either *Y* or *MD*. Persons 2 and 3 both owned *MD* after *OMS*, but only person 3 owned *MD* after *MB*. Hence, Miss Cook is person 3 and Mrs. Dan is person 2. The complete table is now:

	Mr. Au	Mr. Baj	Mrs. Dan	Miss Cook
Owned at start	*MD*	*Y*	*OMS*	*MB*
Bought in 1993	*MB*	*MD*	*Y*	*OMS*
Bought in 1994	*OMS*	*MB*	*MD*	*Y*
Bought in 1995	*Y*	*OMS*	*MB*	*MD*

BOOK PURCHASES

The Janitor and the Principal were chatting as nine customers paid for their purchases. As the group filtered out of the hall, the Alumni Director came up to join her two companions.

"Those nine seemed to find something to their liking," she remarked.

"Yes," responded the Janitor, "but none of them bought more than three books."

"You know," observed the Principal, "it is remarkable how their tastes overlapped. If you took any three out of the nine, there were two of them who bought the same title."

"From which," broke in the Alumni Director brightly, "it can be seen that there was one title purchased by at least three of the buyers."

The Principal arched his eyebrows. However, before he could ask for an explanation, the Alumni Director dashed away to attend to some pressing business.

Can you reconstruct how she reasoned to her conclusion?

Comments

In solving this problem, one should keep in mind the Pigeonhole Principle: if we sort a number of items into categories, and the number of items exceeds the number of categories, then some category must receive at least two items.

There are two cases that might occur. It might happen that there are two customers who did not have a book in common. Put these two customers together with a third customer; what can we say about the purchases of the third customer?

The other possibility is that any pair of customers have purchased a book in common. How many possible pairs of customers are there? At most, how many books were purchased by the nine?

Solution

Consider two possibilities:

(a) There are two customers (call them A and B) who did not have a book in common. Together they have purchased at most six books. Each one of the remaining seven customers must have a purchase in common with either A or B.

Using the Pigeonhole Principle (with the books purchased by A and B as categories and the seven other customers as items), two of the seven customers must have purchased the same book as purchased by either A or B; thus, there are three people who purchased this particular book.

(b) Any pair of customers purchased a book in common. Now there are 36 distinct pairs of customers, but at most 27 books purchased by the nine. Again using the Pigeonhole Principle (with the books as categories and pairs of customers as items), there are two distinct pairs of customers who have the same book in common. But two distinct pairs of customers include either three or four people, and the result again follows.

SAMPLING PATRONS

The Alumni Director wanted an estimate of the proportion of College graduates among the patrons at the book sale, so she interviewed a random sample of no more than 25.

"There was at least one graduate," she reported to the Principal, "but fewer than half were graduates. To the nearest whole percent, that works out to ..." She gave a number that the Principal felt was reasonable.

The Janitor, who had overheard the exchange, objected. "I don't know how many people you talked to, but your answer cannot possibly be right."

When the Alumni Director checked her calculations, she found that, indeed, the correct percentage was without question greater than the figure she gave by three. How many people did she interview?

Comments

This problem will require a fair bit of work. How much will depend on how efficiently you can set it up. The possible fractions with the number of College graduates as numerators and the number of people interviewed as denominators are of the form $\frac{p}{q}$, where p is positive integer and q is a number somewhere between 3 and 25 inclusive. Note that work can be saved by noting, for example, that all fractions with denominators 5 or 10 are covered by those with denominators 20. The percentage determined by the fraction $\frac{p}{q}$ is the nearest integer to $100\frac{p}{q}$; we may suppose that, had this number had a fractional part of 0.5, the Alumni Director could have given the nearest lesser or greater integer, so that the percentage could not be given "without question."

Solution

The details of the solution are more or less onerous, depending on the approach taken. Any percentage which is divisible by 4 or 5 can be written as a fraction with denominator 25 or 20, respectively. It is fairly straightforward to see that the fractions $\frac{1}{25}, \frac{1}{24}, \frac{1}{23}, ..., \frac{1}{10}, \frac{1}{9}$, cover all percentages between 4 and 11; and that $\frac{3}{25}, \frac{3}{24}, ..., \frac{3}{14}$ cover all percentages from 12 to 21. At this point, a variety of techniques can be used to get the other percentages, except the elusive 34.

Probably the cleanest way to show that 34% is not possible is to look at fractions just greater than $\frac{1}{3}$. Consider numerators 2, 3, 4, 5, 6, 7, 8 and find the largest denominator yielding a fraction exceeding $\frac{1}{3}$:

$$\frac{2}{5} \ (40\%), \ \frac{3}{8} \ (37\% \text{ or } 38\%), \ \frac{4}{11} \ (36\%),$$
$$\frac{5}{14} \ (36\%), \ \frac{6}{17} \ (35\%), \frac{7}{20} \ (35\%), \ \frac{8}{23} \ (35\%).$$

Any other fraction will be bigger than one of these or else will not exceed $\frac{1}{3}$.

Thus, the Alumni Director gave the percentage as 34%; the correct percentage was 37%. To realise this percentage, we cannot use the numerators 2 and 3. Since $\frac{4}{10}$ is 40%, $\frac{5}{13}$ is about 38%, and $\frac{6}{16}$ could be either 37% or 38%, the numerators 4, 5, and 6 are out. Similarly, 8 can be eliminated and we are left with the fraction $\frac{7}{19}$.

Hence, there were 19 people interviewed of whom 7 were graduates of the College.

NUMBERS WITH DIFFERENT SUMS

Business at the College Book sale was very active, and as the day drew to a close, only ten volumes remained to be sold. The prices marked on them were: 18, 25, 28, 34, 51, 62, 73, 83, 90 and 95 cents. Finally, in walked the Gemini twins, and the Alumni Director was astonished to note that each twin not only made a purchase, but that the total cost of all the books chosen by one was equal to the total cost for those chosen by the other. "It is surely unusual for such a thing to be possible with so few books," exclaimed the Principal.

"Not at all," grumped the Janitor as he whiffled past. "I've got five dollars that says you cannot give me ten numbers, all under 100, for which it is not possible to find two subsets where the numbers in one subset add up to exactly the same sum as the numbers in the other. You needn't use all of the ten numbers; no number can be used in both subsets, and you must consider only subsets with at least one number in them."

"You're on!" shot back the Principal. "I'll bet I can easily find ten numbers under 100 for which all the subsets have different sums."

But after a long and sleepless night, the Principal had to admit defeat. After an extended session in the furnace room, the Janitor was able to convince him that his quest was indeed hopeless. (Can you explain why?) The Janitor bought them both a beer with his winnings.

Discussing the question later with the Alumni Director, the Principal noted, "So we cannot find ten numbers up to a hundred where all the subsets have different sums. I wonder whether the same thing is true if we allow numbers to go up to 200."

"Well," replied the Alumni Director, "I know for sure that you do not need to go beyond 512. We just have to take the number 1 and nine lowest powers of 2:

1 2 4 8 16 32 64 128 256 512.

With this set of numbers, we can get every sum from 1 right up to 1023 inclusive."

"How do you know that?"

"Think of writing numbers to base 2. Instead of using ten digits, we only use two: 0 and 1. So when we get to the number 'two,' we have to write it '10.' The numbers from one to ten written in base two are:

1 10 11 100 101 110 111 1000 1001 1010.

If we take the numeral 1011101 in base two, we can interpret it as

$$1 \times 2^6 + 0 \times 2^5 + 1 \times 2^4 + 1 \times 2^3 + 1 \times 2^2 + 0 \times 2 + 1$$

$$= 64 + 16 + 8 + 4 + 1 = 93.$$

Every number has such a base 2 representation, which shows how it can be expressed as a sum of distinct powers of 2."

The Janitor poked his head in. "I'll bet that you can make the top number smaller. Suppose that we wanted a set of four numbers for which all subsets have different sums. Then, following your idea, we can take powers of 2: {1, 2, 4, 8}. But, if we give a little at the bottom end, we can get something back at the top. {3, 5, 6, 7} will also work. And for five numbers, taking powers of 2 gives us the set {1, 2, 4, 8, 16}. But {6, 9, 11, 12, 13} will also do."

Now, dear reader, it is over to you. How small can you make the largest of ten positive whole numbers for which all subsets have different sums?

Comments

To show that any collection of ten whole numbers not exceeding 100 must have two subsets with the same sum requires a mathematical statement known as the Pigeonhole Principle. This is a simple-looking but powerful assertion that I hope will draw your ready assent:

> Suppose you have a number of objects that have to be sorted into boxes. If there are more objects than boxes, then some box must have at least two objects in it.

This applies to any possible way of distributing the objects into the boxes. Some boxes might not get any objects at all; others may get only one. But, for sure, some box will get at least two.

Let us apply this principle here. Suppose that we take any set of ten or fewer numbers that do not exceed 100. What can we say about the sum? Well, no number is bigger than 100 and there are no more than 10, so the sum does not exceed 1000. Now let us figure out how many possible subsets we can make of ten numbers. In forming a subset, we have a choice for each number in the set—take it or leave it. Over the whole ten numbers, how many options does this give? Now, imagine boxes numbered 1, 2, 3, ... For each subset, put it in the box marked with its sum. Why must some box receive two sets?

For the problem of finding a minimal set with 10 numbers and all subset sums different, it is a good idea to work your way up. Try first four numbers, five numbers, six numbers, and so on. You will find that you make a fair bit of progress.

Solution

Here is the argument that any set of ten numbers, none exceeding 100, has two subsets with the same sum. The sums of the numbers in any subset cannot exceed 1000. However, there are 1023 subsets containing at least one number. (For the first number, we have two choices as to whether to put it in the subset; having dealt with this number, we have two choices for the second number, for a total of 2^2 choices for the two together; for each of these 4 choices, we have two choices for the third number; and so on. Altogether, the number of choices is $2^{10} = 1024$. But this includes the possibility that no number is taken, so we have to subtract 1.) Since we have more subsets than possible sums, there must be two subsets with the same sum. If these two subsets overlap, remove from each the numbers common to both to get two non-overlapping subsets with the same sum.

As far as I know, an algorithm which will produce a given number of positive whole numbers with the largest as small as possible for which all subset sums are distinct has not been found. Here is an algorithm from the English mathematician R.C. Lyness which is pretty good. Consider the following table:

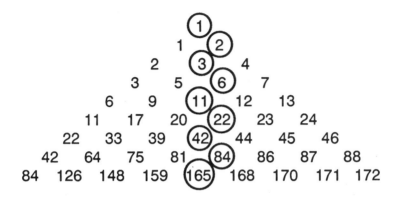

Each row gives a set of numbers of the required type. It is formed from its predecessor as follows: the circled number in each row is either the middle number or the number just to the right of the middle. The first element of each row is the circled number in the preceding row. Add this number to the numbers in the preceeding row to get the other numbers in the row.

From Lyness' algorithm, we find that a suitable set of ten numbers consists of 165, 249, 291, 313, 324, 330, 333, 335, 336, 337.

But Lyness' algorithm is not optimal. For example, in this set of eight numbers the largest entry is only 84 rather than 88:

$$\{20, 40, 71, 77, 80, 82, 83, 84\}$$

(F.V.H.)

Here is a more systematic approach that yields better results:

To find a set of three numbers (trivial, but please bear with me), begin with *four* numbers for which the sums *with the same number of components* all differ, and with the smallest possible maximum number:

e.g., {1, 2, 3, 4} is no good since 4 + 1 = 2 + 3;

but {1, 2, 3, 5} is good (the fact that 5 + 1 = 1+2+3 is not relevant since 5 + 1 has two components and 1 + 2 + 3 has three).

Now starting with {1, 2, 3, 5}, subtract each number from the largest producing {2, 3, 4}, the desired result.

Similarly, to find a set of four numbers, start with {1, 2, 3, 5}; then 6 is no good (6 + 1 = 5 + 2), 7 is no good (7 + 1 = 5 + 3), 8 is good, so take {1, 2, 3, 5, 8}. Now subtract from 8 giving {3, 5, 6, 7}, the desired result. Checking: one number sums are 3 to 7, two number sums from 8 to 13, three number sums from 14 to 18, so no overlap; hence no duplicate sums of any kind exist.

To find a set of five: {1, 2, 3, 5, 8, 14} yields {6, 9, 11, 12, 13}.

To find a set of six: {1, 2, 3, 5, 8, 14, 25} yields {11, 17, 20, 22, 23, 24}.

To find a set of seven: {1, 2, 3, 5, 8, 14, 25, 45} yields {20, 31, 37, 40, 42, 43, 44}.

By now, checking requires a bit more care, because the sum of three numbers can reach 129, and the sum of four numbers is at least 128.

They overlap. However, the sums of three are 129, 127, and less; the sums of four are 128, 130, and more, so there is no duplication.

To find a set of eight: {1, 2, 3, 5, 8, 14, 25, 45, 85} yields {40, 60, 71, 77, 80, 82, 83, 84}.

To find a set of nine:

{1, 2, 3, 5, 8, 14, 25, 45, 85, 162} yields
{77, 117, 137, 148, 154, 157, 159, 160, 161}

(J.R.B.)

In 1987, W.W. Sawyer of Cambridge, England, wrote to me that Richard K. Guy and John Conway had a construction that seemed to work in providing minimal sets of numbers. We first construct a sequence u_m which is to be the largest number for a set of m. The remaining elements of the set are given by

$$a_1 = u_m - u_{m-1}, a_2 = u_m - u_{m-2}, a_3 = u_m - u_{m-3}, \ldots,$$
$$a_m = u_m - u_0$$

Begin by setting $u_0 = 0$ and $u_1 = 1$. For $n \geq 1$, $u_{n+1} = 2u_n - u_{n-r}$, where r is the integer nearest to $\sqrt{2n}$; r can also be described as the unique number which satisfies

$$\frac{r(r-1)}{2} + 1 \leq n \leq \frac{r(r+1)}{2}$$

The following table will give you the numerical details:

n	r	$n-r$	u_n	$a_1, ..., a_n$
1	1	0	1	1
2	2	0	2	1, 2
3	2	1	4	2, 3, 4
4	3	1	7	3, 5, 6, 7
5	3	2	13	6, 9, 11, 12, 13
6	3	3	24	11, 17, 20, 22, 23, 24
7	4	3	44	20, 31, 37, 40, 42, 43, 44
8	4	4	84	40, 60, 71, 77, 80, 82, 83, 84
9	4	5	161	77, 117, 137, 148, 154, 157, 159, 160, 161
10	4	6	309	148, 225, 265, 285, 296, 302, 305, 307, 308, 309
11	5	6	594	285. 433, 510, 550, 570, 581, 587, 590, 592, 593, 594

OVERLAPPING CUSTOMERS

The Principal came into the hall where the book sale was being held as the last customer was leaving. "Well, how did it go?" he asked the Alumni Director.

"Great," she responded. "We had well over 200 people. And I'll tell you something interesting.

"Even though each one came only once, among any three of them, there were two who were in the hall at the same time."

"That is indeed remarkable," marvelled the Principal.

But the Janitor, who had been tuning into this exchange, now interrupted. "I'll tell you something even more remarkable. There were two particular moments such that the time that each customer was present overlapped at least one of the two moments."

"How did you know that?" enquired the Alumni Director. "I did not see you around here all day."

"Elementary, my dear lady. It's all in the little grey cells."

The Principal suddenly emerged from a session of sweet silent thought. "By Will !" he exclaimed. "I do believe that he is right."

Explain how the Janitor might have reasoned his way to his conclusion.

Comments

The number 200 is not material to the problem. With the specific condition, the result would be true regardless of the number of people present. Thus, you might simplify the problem by considering the situation when only three or four people showed up. You might try to model the situation with a diagram in which you mark with segments above a time line the periods when the individuals were present

You should try to describe the moments in terms of key people. Consider for example the first or last to arrive or leave.

Solution

Suppose that Alpha was the first person to depart and Omega the last to arrive. There are two possibilities. Suppose that Omega arrived before Alpha's departure. Then everyone else, being present before Omega arrived and Alpha departed, was present over the whole period from Omega's arrival to Alpha's departure.

On the other hand, suppose that Omega arrived after Alpha left. Consider any third person, Gamma. Let us examine the trio Alpha, Gamma, and Omega. Since Alpha and Omega were not present together, either Alpha and Gamma overlapped or Gamma and Omega overlapped. If Alpha and Gamma overlapped, then, because Gamma left no earlier than Alpha, Gamma must have been present when Alpha departed. On the other hand, if Gamma and Omega overlapped, then Gamma arrived no later than Omega, with the result that Gamma was present when Omega arrived.

Thus, we see that, in any case, any person in the room must have been present when Alpha departed or when Omega arrived.

WINDOW REFLECTIONS

The Principal was admiring the leaded-glass windows of the great hall in which the book sale was held. They exhibited a diagonal lattice pattern, four units across and five down. His finger traced in the air a broken diagonal path beginning in the upper left corner and following the lines, changing direction only at the edge of a pane, much like a billiard ball banking off one cushion onto another.

Watching closely was the Janitor, who noted how difficult it was for the finger to keep on track. "The path will end in the upper right corner," he muttered. "You must have done it yourself," suggested the Alumni Direc-tor. "Oh, no, there is a system," responded the Janitor. "For example, if the numbers of units across and down are both odd, then you would end up at the lower right cor-ner."

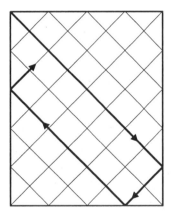

"Well, I was at the Loyal Pacifist Club Ball last night," said the Alumni Director, "and there was a great big window with same pattern 18 units across and 33 down. Where would a path beginning in the upper left corner end up?"

The Janitor's response was almost instantaneous. How would you answer the Alumni Director's question as effortlessly as possible?

Comments

There are a few simple observations that will guide us towards the answer. Observe that the path cannot retrace itself in either direction. If it touches the boundary in the interior of an edge, it continues on in another direction. Therefore the path must end up in another corner—it cannot return to where it started.

Second, suppose the window has a units across and b units down, and suppose that a and b have a common divisor d, so that $a = du$ and $b = dv$ for some integers u and v. Then, for a path beginning in the upper left corner, the destination corner is the same for both the $a \times b$ and the $u \times v$ windows. For example, a 4×6 window is essentially the same as a 2×3 window (try it). In place of the 18×33 window as the LPC Ball, one might as well consider a 6×11 window.

Thus, there are essentially two cases to consider: the window has an odd number of units in each direction; the window has an odd number of units in one direction and an even number in the other.

One way to get an effective handle on the situation is to imagine a square grid superimposed over the window and determined by the contact points of the lead with the edge. For the 4×5 window, this looks like the figure shown at the right.

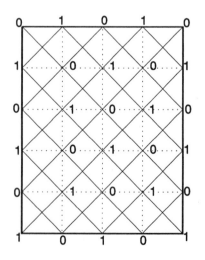

Starting at the upper left hand corner, alternately number the vertices of the superimposed grid 0 and 1. Any path starting in this corner will pass through points marked 0. If the window has an odd number of units in both directions, the path will finish in the diagonally opposite corner. If the window has an odd and an even number of units, it will end up in the adjacent corner in the even direction.

Other Questions

Given the dimensions a and b, investigate the number of distinct paths that are possible when you are allowed to start tracing anywhere in the window.

One might also consider the type of leaded window in which the lead does not emanate from any corner. In this case, each path returns to its starting point. Are there any dimensions for which a single path traverses all the leaded lines before returning to its starting point?

TOY FROM THE BEYOND

Like many old buildings, the College has a ghost. There are many accounts of the appearance of this wraith, Ivan, particularly to students who used to be resident in the College. On one occasion, a student was discussing celestial toys with him.

"Ivan, what is this frisball that, according to you, children on the Other Side enjoy playing with so much?"

The ghost put his mug of beer on the table, strode to the blackboard and drew a disc (Figure 1). "You sew together diagonally opposite points of a disc of material: *A* to *A*, *B* to *B*, and so on."

"Like a football?"

"No. Quite different. You cannot make it here, because you have only three dimensions. But I can show you one."

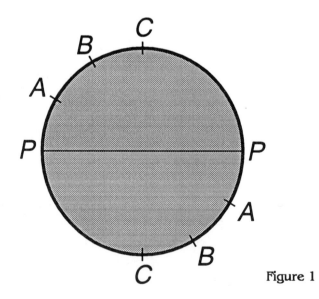

Figure 1

He withdrew an object from under his cloak. The space around it shimmered, and the student had a peculiar sense of looking into the fourth dimension. He held it up by a string around its middle (indicated in Figure 1).

"Unlike for a football, there is no way in which this string can slide off. Funnily enough, if I were to wrap the string around twice instead of just once (to pass through *P* and *Q* in Figure 2), it would slip right off. Let me demonstrate."

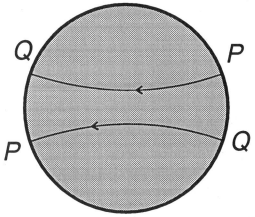

Figure 2

Unfortunately, at that moment, a shaft of dawn light shone through the student's window, and the spectre vanished along with his frisball.

Comments

What the ghost had was a geometrical object known to mathematicians as a projective plane. While indeed it cannot be realized in ordinary three-dimensional space, if one could add a fourth dimension, introducing a direction which is somehow perpendicular not only to two independent horizontal directions but also to a vertical direction, then in this four dimensional space, it would be possible to construct a "frisball."

To understand how this might be done, consider how to construct a torus (the idealization of the shape of a doughnut or bagel). Begin with a pliable and stretchable rectangle (Figure 3). Gluing one pair of opposite sides together (Figure 4) in the same direction would give an open cylinder. Gluing together the other pair of opposite sides (the ends of the cylinder), would then give us a torus (Figure 5).

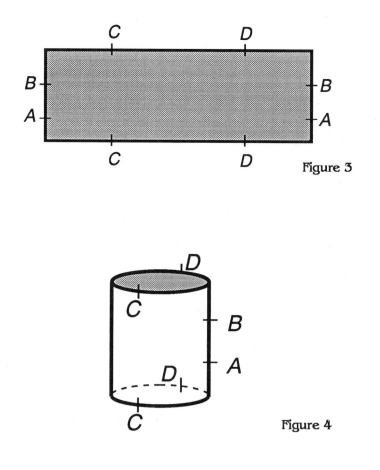

Figure 3

Figure 4

Figure 5

Now begin with the rectangle again. This time, glue one pair of opposite sides together after twisting one end of the rectangle around 180° (Figure 6); this yields a one-sided object called a Möbius band. To understand that this object is a little stranger than an open cylinder, cut it lengthwise down the centre. What happens is neatly described in a bit of verse:

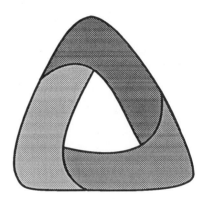

Figure 6

A mathematician once confided
That a Möbius band is one-sided;
And you'll get quite a laugh
When you cut one in half,
For it stays in one piece when divided.

Now (you will have to imagine this—you won't be able to actually do it), glue the other two sides of the rectangle together, again after giving one a 180° twist. What you now get is a projective plane. If you had not made the twist before gluing the second pair of sides together, you would have got another impossible object called a Klein bottle.

Solution

The ghost returned to the student the following evening and explained how to slip the string off the frisball. "*P* and *Q* are points on the string which has been twice wrapped around the frisball. Let me tighten it to make the loop smaller, through positions *RS*, then *TU*, until you can clearly see how the string is sitting on the surface of the frisball and comes right off" (Figure 7).

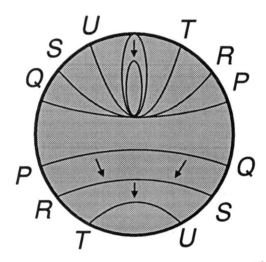

Figure 7

Appendix

Casting Out Nines

Suppose we begin with a whole number n written out in base 10. Add up its digits to get a second number. If this number is greater than 9, sum its digits to get a third number. Repeat the process until you obtain a single digit. We will call this the digital sum of n and denote it by $d(n)$.

For example,

$$d(1583776) = d(37) = d(10) = 1$$

A number is divisible by 9 if and only if its digital sum is 9. If the number is not divisible by 9, then its digital sum is the remainder upon division by 9. Thus, 2387 has digital sum 2 and, indeed $2387 = 9 \times 265 + 2$.

This property depends on the fact that the difference between any number

$$n = a_0 + 10a_1 + 100a_2 + 1000a_3 + \cdots$$

and the sum $a_0 + a_1 + a_2 + a_3 + \cdots$ of its digits is a multiple of 9, namely $9a_1 + 99a_2 + 999a_3 + \cdots$.

We can use "casting out nines" as a rough check on addition and multiplication. If m and n are two whole numbers, then $d(m+n)$ is the digital sum of

$d(m) + d(n)$ and $d(m \times n)$ is the digital sum of $d(m) \times d(n)$. For example,

$d(372 + 467) = d(839) = d(20) = 2$

$d(372) + d(467) = 3 + 8 = 11;\ d(11) = 2$

$d(372 \times 467) = d(173724) = d(24) = 6$

$d(372) \times d(467) = 3 \times 8 = 24;\ d(24) = 6$

Thus if we need the digital sum of the sum or product of two numbers, we need deal only with the individual digital sums and apply the desired operation to them.

The Pigeonhole Principle

The Pigeonhole Principle states that if you have a number of objects to be sorted into categories, and there are more objects than categories, then some category must have at least two objects belonging to it.

Key to Contributors

Certain individuals contributed ideas used in this book and are acknowledged throughout the text by their initials. Many of these responded regularly to the "Aftermath" column in the *University of Toronto Alumni Magazine*. I am grateful for the contributions of the following:

DCB	Donald C. Baillie, Toronto, Ontario
GB	George C. Baker, Kentville, Nova Scotia
JRB	John R. Bird, Toronto, Ontario
EC	Ellen Carlisle, Toronto, Ontario

RC	Rebecca Chan, University of Toronto student
RCh	Robert Cherniak, Toronto, Ontario
MC	Mr. Chow, University of Toronto student
GAC	G. C. Cooper, Thornhill, Ontario
AC	Alan Craig, Brampton, Ontario
WSAD	W. S. A. Dale, London, Ontario
SD	Sallianne Dech, Sarnia, Ontario
AF	Allan Farnell, Toronto, Ontario
JPF	J. P. Fedorkiw, Etobicoke, Ontario
JGF	John G. Flatman, Timmis, Ontario
ASG	Arthur S. G. Grant, Mahone Bay, Nova Scotia
JG-McL	John Grant-McLoughlin, Corner Brook, Newfoundland
GH	George Harrap, Agincourt, Ontario
FVH	Fred V. Harrison, Etobicoke, Ontario
LGH	Lloyd G. Hinton, Aurora, Ontario
WK	Wieslaw Karpinski, Clifford, Ontario
JSM	J. S. Martin, Edmonton, Alberta
RCM	R. C. Martin, Toronto, Ontario
FHR	F. H. Rooke, Toronto, Ontario
WWS	W. W. Sawyer, Cambridge, England
DS	Doug Sutherland, London, Ontario
LJU	Leslie J. Upton, Mississauga, Ontario
AWW	A. W. Wallace, Toronto, Ontario

Index